# My Dog's Mystery Adventure

## Also by Karen B. London

*Treat Everyone Like a Dog: How a Dog Trainer's World View Can Improve Your Life*

*Cows, Ants, Termites, and Me: Revealing the World of Animals One Newspaper Column at a Time*

## By Karen B. London (co-authored with Patricia B. McConnell)

*Feeling Outnumbered?: How to Manage and Enjoy Your Multi-Dog Household*

*Feisty Fido: Help for the Leash-Reactive Dog*

*Way to Go! How to Housetrain a Dog of Any Age*

*Play Together, Stay Together: Happy and Healthy Play Between People and Dogs*

*Love Has No Age Limit: Welcoming an Adopted Dog Into Your Home*

# My Dog's Mystery Adventure

**And Other Stories from a Canine Behaviorist and Dog Trainer**

Karen B. London, PhD

Animal Point Press

Karen B. London/Animal Point Press
3100 West Shannon Drive
Flagstaff, Arizona 86001

Cover design by Heather Brown of Cultural Sponge

E-book: ISBN 978--1-952960-05-5
Paperback: ISBN 978-952960-04-8

My Dog's Mystery Adventure: And Other Stories from a Canine Behaviorist and Dog Trainer / Karen B. London. -- 1st ed.

*To Patricia B. McConnell, PhD, who changed my life when she began teaching me about dogs in 1997 and continues to be my mentor even now. Her wisdom and humor are gifts I can never repay.*

# Table of Contents

# Introduction

In September of 2008, I wrote a guest piece for my local paper, the *Arizona Daily Sun*. It appeared in the *Animal Instincts* column, which was renamed *The London Zoo* when I took over the column in December of that year. Obviously, my work has nothing to do with the *actual* London Zoo, and everything to do with my last name and my love of animals. Growing up, we had a sign on one of our bookshelves with the words "The London Library" on it and I used to think it would be fun to have so many pets that we needed a sign for a zoo named after us as well.

The column was originally a pet column that I morphed into a column covering far more of the biodiversity in the animal world than our pets. I have written about all kinds of animals—from banana slugs and llamas to falcons and honey pot ants. Despite the diversity of my subject matter, there is one species I have written about far more than any other species, and that's the domestic dog, *Canus lupus familiaris*. That's because I love them, I know them well, and I find them fascinating. It's an exciting time to be writing about dogs because there is greater interest in them than ever. Scientific studies about them are increasingly common, and it is quite mainstream to recognize their value in our lives as family members and best friends.

Out of all the animals in the world, I am most familiar with our best friends, the dogs, so I never run out of canine topics to cover. Still, I was quite surprised to realize how many columns have been about our canine buddies. Inside, you will find 50 columns from September 2008 through November 2021. My hope is that these collected columns will appeal to your heart and your mind, just as dogs themselves do.

# Author's Note

I keep learning, so not all that's written here reflects what I now believe and know, or how I behave currently. As a collection of columns written over many years, this book reflects my views and knowledge over time and not at a particular moment.

Throughout the over 16 years I have been writing the animal column, I have been anything but consistent on the grammatical front. There have been multiple editors at the paper handling my column, and I have also changed along with the times.

I used to use all kinds of language gymnastics to keep things gender neutral rather than always go with "he" or "she." As the years have gone by, I have altered the pronouns I use for dogs in a manner that parallels the use of pronouns in the rest of society. Now I just use "they" for individual dogs as well as for groups of dogs. This change reflects the more conscious, respectful, and thoughtful use of pronouns in society in general and I embrace it.

My irregular use of who/that/which when referring to individual animals and to species is harder to explain, but I am aware that the proper use of these relative pronouns is not something I excel at. I use both metric and imperial measurements because I'm comfortable with both and it didn't occur to me until rather recently that I should stick with just one. (I've subsequently chosen the metric system simply because that's what most of the world and the entire scientific community uses.)

The columns have been edited for uniformity in a general sense, but far more important was sharing them in what was essentially their original form. Writing the columns and assembling them into a book has been a process emphasizing flexibility over consistency. If grammar inconsistencies catch your attention, consider the opportunities to find them within these pages a fun challenge. If they don't catch your attention, then I have no more explaining to do.

# What's in a Name Beyond Fido and Spot?

***September 22, 2008***

Ever since I learned that John von Neumann, the most famous mathematician of the 20th century, named his dog Inverse, I have been fascinated by the names we choose for our dogs. Dog names seem to provide people an opportunity to express their views, hobbies, philosophies, and general perspective.

When President Garfield named his dog Veto, he certainly had something very different to say to the world than what singer Jimmy Buffet was conveying with his choice of Cheeseburger for his canine companion. As a rule, presidents' dogs have had some pretty colorful names, perhaps because of the importance of dogs in the lives of presidents. As Harry Truman said, "If you want a friend in Washington, get a dog." Lyndon Johnson, true to his no-nonsense personality, named his two Beagles Him and Her. FDR chose the moniker President for his Great Dane. Gerald Ford's Golden Retriever expressed a value of tremendous importance with the name of Liberty, whereas Jimmy Carter's mixed breed had the folksier title of Grits. Similarly, George Patton's dog Willy (named after William the Conqueror) and George Washington's dogs Drunkard, Taster, and Tipsy illustrate the full spectrum of naming motivations among generals.

In Wisconsin, I knew many dogs named Lambeau, Favre, Lombardi, and Packer, showing the widespread reverence for the Green Bay Packers in the Cheesehead State. Elsewhere in sports, hordes of golfers have dogs named Putter, Divot, Caddy, or Tiger. One family took the golf theme more seriously, conveying bathroom-related information to other members of the household with comments such as "Divot already bogeyed and double-bogeyed this morning." I know runners with dogs named Marathon and Sprint and a badminton player whose Pointer is called Birdie.

Sometimes it's not sports, but a person's background that is the source of their dogs' names, most notably Adam Sandler, whose Bulldog is named Matzoh Ball. I have a friend very proud of his Irish heritage who named his Irish Water Spaniel Shamrock, and a colleague whose choice of Uffda reflected a similar pride in her Norwegian background.

My husband and I gave our half black Lab, half handsome stranger the name Bugsy, as a tribute to our research interests—I did my dissertation on the defensive behavior of tropical social wasps, and his was on the population dynamics of bark beetles, mites, and fungi. In a similar way, Sheila Madrak, the *Animal Instincts* columnist, drew on her love of sea turtles to choose the name Etta for one of her dogs. The loggerhead sea turtle's scientific name is *Caretta caretta*.

Paris Hilton and Jake Gyllenhaal both chose names based on books. If you can't decide which celebrity is more intellectual based on their choices, then it's time for a remedial course in the literary arts. Jake's dogs are Atticus and Boo Radley while Paris' dogs are Tinkerbell and Bambi.

As a canine behaviorist, dogs' names provide clues to me about their history and behavior. Dogs named Ditch and Freeway were most often found abandoned as roadway strays. Dogs named Solo and Uno are often from singleton litters, which can result in an enormous range of predictable behavioral issues. I tend to suspect that dogs named Baby are being coddled, and that owners who choose to name their dogs Spike and Killer often find aggressive behavior more amusing than problematic. I am sometimes very wrong in my assumptions. One Spike turned out to belong to a beach volleyball fanatic, so I always investigate hunches based on names before acting on them.

It's unpredictable which names will become common. Few dogs have been named Martha after the Beatles' song *Martha My Dear* but the real inspiration for that song was Paul McCartney's own Old English Sheepdog named, you guessed it, Martha. Interestingly, both the actress Shirley McClaine and King Edward the VII of Great Britain named their dogs Caesar. Simba has been popular since Disney's *The Lion King* premiered, and Boomerang is a favorite for fetching dogs.

Whatever you choose to name your dog, the choice will probably tell people more about you than about your dog. With that in mind, I am waiting for the day when I notice the most common dog names are Happy, Patience, Tolerance, and Joy.

**I contributed this as a guest to Sheila Madrak's *Animal Instincts* column, but I became the animal columnist less than 3 months later.**

## Canine's Tail Tells a Tale

### *December 29, 2008*

"I can't believe she bit me! She was wagging her tail!"

I've heard this exclamation of surprise countless times, and it's a result of a common misconception about canine behavior—that a tail wag is a sure sign of friendliness. The most accurate interpretation of a tail wag is *not* that the dog is friendly, but rather that the dog is willing to interact. Friendly dogs, being willing to interact, wag their tails. However, there are unfriendly ways of interacting, and dogs with these intentions also wag their tails. The details of a tail wag reveal its meaning.

Generally, the closer the start of the wagging movement is to the head, the more likely it is to indicate friendliness. The full body wag that extends from the shoulders through the belly to the hips and the tail is the classic friendly tail wag, performed by some of the world's most affable creatures. The tail wagging that includes the hips in full swing as well as an active tail is also most likely a sign of high sociability. These "tail" wags can even be recognized in individuals without tails. When just the tail moves, the wag may or may not have anything to do with being sociable. When the only part of the tail that moves is the tip, the dog is most likely not friendly.

The speed of the tail wag is also very informative, with faster tail wags indicating friendliness. A slow tail wag is actually a sign of an absence of friendliness. Sometimes a dog who is a bit nervous or even reserved may wag slowly, which simply shows some hesitancy about interacting. Other times, a slow tail wag may signify aggressive tendencies.

Another revealing attribute of a wagging tail is its tenseness. A stiff tail wag is generally not a good sign. The more relaxed a tail is while it is wagging, the more likely it is that the tail wag shows friendliness. A flexible, fluid motion in the tail looks friendly, and has a very different appearance and meaning than a rigid tail.

Sometimes a dog wags the tail in a rotating motion starting from the base of the tail with the tip of the tail tracing a broad circle. The "circle wag," as it is called, is one of the positive signals that experienced dog handlers attend to when evaluating dogs because it is generally seen only in dogs in a friendly mood.

To sum up, the friendliest type of tail wag is the full body wag with a rapidly moving, flexible tail. In contrast, the least friendly tail wag, and the one that is of most concern in terms of potential aggression, is the slow twitch of only the tip of the tail on an otherwise unmoving, rigid tail, attached to an equally stiff body. Only careful attention to the specifics of a tail wag can discern its significance. The meaning of tail wags is far more complicated than the simple but misleading idea that a wagging tail indicates a friendly dog.

**There are so many things I wish people knew about dogs, but the fact that a wagging tail is not a sure indication that a dog is friendly tops the list.**

## The Movie *Bolt* Gets a Lot Right

***January 26, 2009***

It's always fun to scoff at Hollywood's attempts to be scientifically correct, and the opportunities come up so often that resisting the urge to poke fun is usually futile. That's why it was such a surprise to catch Disney getting so much right in the recent animated movie *Bolt*. There were three particularly charming scenes that got a lot of canine behavior spot-on (sorry!).

In one scene, the dog Bolt plays with another dog, and the behavior patterns that he exhibits are accurate. He and his playmate both perform play bows at the start of play. This behavior pattern consists of putting their elbows on the ground while leaving their back end up. As is often the case in real life, the dogs look at each other with the classic mammalian play face of an open mouth and a relaxed face. They wag their tails, too, which is also commonly seen when dogs perform play bows. Then, after the first set of play bows, there is a pause before a game of chase begins, then another break in the play before it gets going again. The behavior sequence is a textbook example of play between unfamiliar dogs, especially in regards to the presence of so many play bows.

Another scene shows Bolt as a puppy completely obsessed with a squeaky carrot toy. He squeaks it repeatedly, and also pounces on it, grabs it, and shakes it, just as many real-life toy-motivated dogs do. His behavior reveals the same big-footed, clumsy goofiness so typical in real-life young dogs. This toy remains a favorite of his for years, which is also remarkably common in the real world.

The third refreshingly accurate scene in *Bolt* involves a streetwise cat named Mittens teaching the sheltered Bolt how to beg for food from people. Mittens is very specific and quite savvy about how to look as dear as possible in order to get humans to relinquish their food. Mittens' instructions to Bolt include cocking his head, opening his eyes wider as he tilts his head forward, putting one ear up and one ear down, whimpering, and lifting his paw.

The visuals of Bolt performing each step according to the cat's instructions make for a hilarious montage as Bolt's body postures and facial expressions combine in ever-more-effective ways for getting people to say "Awww" and surrender their food.

In another toast to the reality of life, when this scrawny cat attempts the exact same behaviors to beg for food, the people tell her to scram or slam the door on her. Cats are generally less effective at getting people to give them food. It seems no animal can churn up humans' sympathetic giving natures like the dog.

*Bolt* is no nature documentary. As in most animated films, the animals talk, plot elaborate schemes, and display all the other human-inspired behavior you'd expect from an action flick. Nonetheless, when the animals were being true to their species, all I could think was, "Kudos to Hollywood for this rare and unexpected accuracy!"

**So many animal movies are ruined for me, and I'm eternally grateful that *Bolt* isn't one of them.**

## An Accidental Punishment

***March 9, 2009***

The fall was so horrendous that while I was in the air, I saw my feet silhouetted against the night sky and I had time to think, "This is really going to hurt when I hit," before I smacked the icy ground. I avoided personal injury, but the real damage was to the unusually-named Who, a sweet English Springer Spaniel puppy whose leash I inadvertently jerked as I fell, and the timing was disastrous.

Who had just started to squat to urinate when I lost my balance. In dog training terms, she received a punishment for urinating, in the form of my accidentally yanking her leash. Reputable dog trainers these days use positive reinforcement, which has many advantages over using punishment, including superior training of dogs plus better relationships between people and dogs.

Punishment-oriented training techniques can result in great damage, especially when errors are made. One treat given at the wrong time may confuse your dog, but rarely does it have such a powerful adverse effect on either the training or on the relationship. Conversely, one wrongly timed punishment can do tremendous harm.

If I had been trying to teach Who not to relieve herself while on a leash, the timing of my fall would have been excellent. The strong punishment of having her leash jerked at the exact moment that she initiated the action made her less likely to perform that particular behavior in the future. The problem was, I had spent many days trying to get her to urinate while on a leash, and this single incident was a terrible setback requiring weeks to overcome.

The setback was particularly unfortunate because Who was a real challenge to house train. Most puppies are relatively easy to house train. I say "relatively easy" because even in the best of circumstances, there are generally interruptions to the normal sleep pattern, the occasional cleanup, and some frustration in both dogs and people.

Unlike most puppies, Who would not eliminate when she was on leash, but she would go when she was in her crate. Off leash, she would go, but since it wasn't always safe to have her running free, she needed to learn to eliminate even when on leash. Hence, I was walking around with her on leash on our 150-acre farm late at night when it was below zero even though she would have been perfectly safe (and willing to urinate!) without the leash.

I was prepared to give her an entire pocket full of liver biscotti once she went in order to help her understand what I wanted her to do. Because my plan went awry, she was instead punished for the very action I was hoping to positively reinforce.

I was lucky Who was not seriously injured, and that my pride was more hurt than any body parts, but the damage to our relationship and the damage to my training efforts were upsetting. Punishment is the least forgiving of all training techniques, which is a great reason not to use it, especially if you have the misfortune to be house training a puppy during February in Wisconsin.

**Training mistakes are often informative as well as instructive, and all trainers have piles of stories of errors and unfortunate events. This was one of the stories I told Hannah Branigan as a guest on her podcast *Drinking From The Toilet* when the theme was training bloopers.**

## Is Your Dog Like the Flintstones' Dino?

***March 30, 2009***

"My dog turns into a lunatic when people come to my house. He jumps up on them, leaps around, and generally makes a pest of himself. In a few minutes, he calms down and returns to being his usual sweet self. How do I get him to act that way from the start?" This is a common question for dog trainers.

Many dogs are what professionals call "effusive greeters." To teach your dog to be more polite with visitors, it is necessary to decide what you would like him to do instead of jumping up on them. Most people want their dogs to sit when greeting people, but perhaps you'd like him to head away from the door while you let the people in. Both behaviors are incompatible with jumping up since a dog cannot simultaneously sit down or head away from the door and jump up on a visitor.

If you want your dog to sit when visitors arrive, teach your dog that sitting results in good things, but that all other behavior causes the visitor to leave. If your dog sits, your visitor should give him a treat. If he jumps up or is otherwise out of control, the visitor should immediately walk out. After a 30-second wait, the visitor should return. Try this for 3 to 5 repetitions. Having a visitor leave has a sobering effect on most friendly dogs because instead of getting to interact, the person disappears.

Most dogs get the hang of the proper behavior during a session, but they still goof the next time somebody shows up. At first, dogs usually need four to five chances to get it right. After a few weeks of daily practice, most dogs get it right after just two or three tries, but they need several more practice sessions before they succeed on the first try. In addition to practicing with visitors, do the same thing yourself each time you come home—give treats to your dog if he sits, but walk out if he jumps up or otherwise acts too exuberantly.

A second approach is especially effective with toy-motivated dogs. Teaching your dog to go get a toy when visitors arrive gets the dog away from the visitors. As soon as a friend enters, wave a toy briefly in front of your dog's face and then toss the toy away from the door. Start with tosses of just a few feet. Gradually start throwing the toy further away. Do this each time a visitor arrives until your dog is in the habit of going to get a toy whenever someone comes over. It also works to have the visitor throw the toy. Eventually, most dogs will go get a toy without you having to throw it because the visitor coming in becomes the cue to go get a toy.

These techniques are based on two important training principles: 1) Focus on what you do want your dog to do, rather than what you don't want him to do, and 2) Teach your dog to perform a behavior that is incompatible with the undesirable behavior.

**Consistency is the key to making either technique truly effective. You'll also have more success if you play with the dog who got a toy, so they are reinforced with fun for behaving as you'd like when visitors arrive.**

# Portuguese Water Dogs 101

*April 20, 2009*

Whenever the public eye turns to a dog of a particular breed, that breed rises in popularity. (I originally typed "pupularity" by mistake, which is fitting.)

Years ago, I had countless clients who got a Jack Russell Terrier after being inspired by Eddie on the TV show *Frasier* but ended up with behavior issues. Many wanted a cute dog to cuddle with or talk to, and instead got a mischievous, relentless, and clever problem-solving dog who had been bred to hunt and to be tenacious—traits that didn't make for a good pet for the average family.

The even more alarming *101 Dalmatians* movie-fueled debacle resulted in many families with young children having dogs who were bred to run all day long alongside carriages. Dalmatians are perfect for people who want to take their dog out for a 50-mile run before breakfast and then don't mind when that same dog asks them what they are going to do all day.

Having survived the crazes created by other breeds, I was thrilled to learn that the Obamas have welcomed a Portuguese Water Dog into their family. They are generally easygoing, social, trainable, athletic, and make friendly companions. This breed requires lots of social time and attention, and needs plenty of exercise to maintain a proper weight. If these needs are met, they generally make lovely family dogs.

Naturally, there will be great interest in Portuguese Water Dogs now that one is living in (and probably occasionally peeing in) the White House. So, here are some basics about the breed.

First of all, one reason the Obamas chose a Portuguese Water Dog is that their elder daughter has allergies. This breed sheds less than most, so many people with allergies can tolerate them. No dog is completely non-allergenic and people with moderate to severe allergies are still likely to react to them.

Portuguese Water Dogs are medium-sized dogs, typically just under 20 inches high and about 45 pounds. They have shiny coats that are most often wavy or loosely curled. They are black, white, or brown, sometimes also having markings or spots of another color. This breed is regularly mistaken for a Poodle, which it only superficially resembles. Folks in the dog world refer to Portuguese Water Dogs as PWDs.

Historically, Portuguese Water Dogs worked on fishing boats off the coast of Portugal assisting fishermen. Great swimmers with webbed feet, these dogs carried messages to other boats, helped people manage the nets in the water and even "herded" fish into nets. They caught fish that escaped the nets and dove deep into the water to retrieve lost items and men who fell overboard. Some used their senses of smell or sight while on the bow of the boat to detect schools of fish and then let the fishermen know fish were nearby. There are even reports of them acting as a sort of foghorn when fog was heavy, barking continuously from the bow of the boat to alert others to the presence of their vessel.

That's it for the crash course on Portuguese Water Dogs. And yes, everything in this article will be on the exam.

**Generally, choosing a dog breed because you saw one member of the breed on TV or in a movie is a risky proposition—just ask the thousands of people who adopted a Border Collie—but not a flock of sheep!—after seeing *Babe*.**

## Surviving the Rabies Quarantine

*April 27, 2009*

If you find yourself without enough to worry about lately, you're in luck. You can now fret about the fact that rabid foxes and skunks have been found in the Flagstaff area, and there have been reports of attacks by such animals. As a result, there is a three-month quarantine on our pets. Dogs and cats must be kept on leash or in enclosed areas.

Trying to keep pets happy and satisfied when they can't run freely outside is challenging. The prospect of living with an under-exercised animal who chews, whines, barks, or scratches can be alarming. The devil really does find work for idle paws.

The key to keeping you or your pets from going crazy during the quarantine is playing games indoors, lots of mental exercise, and taking advantage of any opportunities to be outside.

Training provides mental exercise, plus you reap the benefits of having a better-trained animal. Ask your dog to "Sit," "Stay," "Lie Down," "Wait," or "Come." Cats can also learn new skills through training, and two books to get you started are Miriam Fields-Babineau's *Cat Training in 10 Minutes* and Karen Pryor's *Getting Started: Clicker Training for Cats.*

Tricks can be another diversion. Learning tricks challenges the mind and makes many dogs tired even without physical exercise. (Remember the overwhelming exhaustion of final exams?) Popular dog tricks include crawl, spin, beg, roll over, wave, shake, and high-five. If your dog likes to retrieve and is into toys, teach her the names of her toys so that you can tell her which one to get. Or, teach her to bring you a tissue when you sneeze. Common cat tricks include coming when called, perching on your shoulder, and rolling on their back with all four feet in the air.

Playing games inside can help keep pets entertained. Hide-and-seek, find your treat, chase, fetch, and tug are all great canine games. Cat games often involve chasing toys such as fake mice pulled on a piece of yarn. Some cats like to fetch, but many are happiest playing "soccer" by dribbling little balls around. My favorite cat toy is the Kitty Teaser, which allows cats to chase after a fast-moving item dangled at the end of a fishing-pole like holder. Kitty perches and towers that let them leap and climb up high or sit by a sunny window help satisfy many cats' need for stimulation.

To get outside despite the quarantine, leash walks for dogs are an obvious possibility, but some cats enjoy them, too. Before you take your cat out on a leash, get a harness. (Don't attach a leash to her collar—she's too likely to get hurt or struggle free.) Not all cats will tolerate this confinement, but many will. For dogs, fenced-in yards and the dog park (for social dogs only!) offer outdoor opportunities for fun and exercise.

Boredom is the enemy of the well-behaved pet. As long as your pets are using their minds, whether it's to play or to work, they are getting the mental exercise that helps them be dogs and cats who sleep on the rug in front of the fireplace instead of chewing it up.

**The thought of keeping our pets occupied and happy during a quarantine may be overwhelming, but enrichment is what will save us all from the perils of living with a bored pet.**

## Getting the Most Out of Going to the Dog Park

*June 1, 2009*

There are people who consider dog parks to be the canine equivalent of a gladiator pit, but other people have more fun there than during the first snowfall of the year.

Both viewpoints have merit. Dog parks offer opportunities for dogs such as social play, running off leash, and lots of smells for our dogs' noses and brains to enjoy.

They also present the potential for perils. I recommend following certain guidelines in order to get the most out of going to the dog park.

First, the dog park is not for every dog. It is for well-socialized, happy, friendly, healthy adult and adolescent dogs. The dog park is no place for a young puppy who is likely to get overwhelmed or to catch something contagious. It is not for shy or fearful dogs. It is not for dogs with an injury since the exuberant play at dog parks can cause it to worsen. The dog park is among the very worst places to take an aggressive dog or one who is scared of other dogs.

Second, you should focus on your dog. Pay attention to supervising the dogs, rather than simply socializing with other people. I'm not saying you shouldn't make friends with people at the park and talk with them. Rather, this human activity must be secondary to watching over your dogs as they interact. Learn about dog body language and use what you know to evaluate the situation. If your dog seems overwhelmed or scared, step in and save him. You know your own dog best—pay attention to that inner voice that says something is not okay. Don't let your dog be bullied! Likewise, don't let another owner bully you into accepting a situation for your dog that you don't find acceptable. Beware of anybody who tries to stop you from looking out for your dog's best interests by saying, "He needs to toughen up and learn how to take it," or "That's just how my dog plays—trust me, it's fine." Don't hesitate to leave if it gets out of hand.

Third, I advise you to keep moving. Dogs tend to run around and greet other dogs and then dash off again, possibly smelling the same things, running in parallel, or briefly wrestling with one other dog. The rough-and-tumble wrestle play with piles of dogs is unnatural in most cases, and it oh-so-very-often leads to dogs becoming aggressive or afraid. Many problems can be avoided if you walk around and have your dog explore rather than standing in one place. It's especially critical not to huddle around the entrance and socialize. The majority of serious problems at dog parks occur around the entrance because of over arousal and tension as a result of the crowds and constant comings and goings of dogs in that area.

Dog parks can be joyful places for dogs and humans to play. The more people know what to do, what not to do, and how to supervise and spot trouble, the more safe and healthy fun can be had by all.

**In the best of all possible worlds, dog parks can be wonderful places for socializing and play, but that requires that a large number of people have substantial knowledge about dog behavior. Many dogs have had dangerous and upsetting experiences at dog parks, and there are therefore many valid criticisms of them.**

# Bottoms Up in the World of Behavior

### *June 8, 2009*

At no time when I imagined speaking at the national conference of the Association of Professional Dog Trainers did I envision a picture of me with my derriere up in the air to introduce me to 1,000 of my colleagues. I was to give a talk in an all-day play symposium, and the speaker before me was complimenting me on being one of the most playful people she knows. To illustrate her point, she showed a photograph of me performing a canine play bow to a black Lab whom I was trying to entice to play. To perform a play bow, dogs lower the front end of the body and rest on their elbows with their back end up in the air, so this was the ridiculous position I was in. A friend in attendance later told me that I was an "ass-et" to the organization. And yes, she was laughing when she said it.

It was not the first time in my work as an applied animal behaviorist that I had found myself in this position. Not even counting all the time I have spent looking under furniture in my office for lost dog toys, there is much work that is done in the bottoms-up position. For example, one of the first house calls I went on when I was learning to work with animals with behavioral issues involved house soiling by cats. Many such cases involve cats who do not like the litter box and need a new design in order to help them return to it. Another common source of house soiling is cats who have developed a urinary tract infection and stop using the litter box because they associate the pain of eliminating with that place and refuse to go near it.

This case was not so simple. It involved three or four cats who were forsaking the bathroom for the carpets and closets, and at least one who was spraying the walls. Such spraying of vertical surfaces indicates marking behavior and the potential for serious social issues going on rather than simple litter box problems. Figuring out all the places the cats had urinated was not easy. Usually, you can track the smell and localize it to just one place, but this whole house was strongly scented with eau de cat. The first step was to figure out where the incidents were happening and then videotape those spots to figure out which cat was doing what and where.

We began phase one of our analysis of the situation—sniffing out urine marks. This meant crawling around the house sniffing the walls and floors. Prior to this time, I had not realized that my professional life was going to involve so much time spent on all fours with my head down and my back end up in the air. As my mentor, Patricia (Trisha) B. McConnell, PhD said as we were leaving the house, "Do you think we can advertise for our profession with, 'You, too, can get a PhD and spend time crawling around, fanny up, while you sniff for cat pee!'?"

**After more than a quarter century, I still love my work and I still regularly find myself in ridiculous postures in the course of doing it. Incidentally, it was also**

**Trisha McConnell who spoke before me at the play symposium and introduced me with the play bow photo.**

## Domestic Dog has Influenced Wolf Coloring

***June 15, 2009***

Dogs and wolves have been considered, at various times, to be different but closely related species and to be two subspecies of the same species. In any case, it is well documented that dogs and wolves can breed with one another.

New evidence suggests that former breeding with the domestic dog may actually be helping populations of wild wolves to flourish. According to a study by Tovi M. Anderson and 14 co-authors, the genetic diversity related to coat colors that has been preserved in the domestic dog may be helping wolves to survive. Their paper appears in the March 6, 2009, issue of the prestigious journal *Science* (Volume 323, Pages 1339–1343).

These researchers observed that the generally rare black color form of the wolf was more common in certain habitats than in others. This dark color is extremely rare among wolves living in the high tundra and who follow caribou seasonally. Black individuals are most common in populations of wolves who live all year in boreal forests and who hunt prey that do not migrate. There is actually a gradual change in the prevalence of black wolves across their range, so that in the areas between the high tundra and the forests, there is an intermediate proportion of wolves that are black.

It has been proposed that the advantage of the dark coloration for wolves that live in the forest may be that it helps them to remain concealed when they hunt. Coat colors and habitat often correspond, and it is often assumed that this is a result of natural selection. Interestingly, the genetics behind this dark color form in wolves are different than the most common genetic sources of color variation in many animals. Usually, coloration involves very complex genetics. In wolves the genetics that relate to the dark coloration are relatively simple. In fact, gray and black coat colors sort as a Mendelian trait. This means, basically, that if an individual gets a "gray" gene from each parent, then that individual will have a gray coat, but if an individual receives a "black" gene from one parent, the coat color will be black. (Receiving two "black" genes, one from each parent, also results in a black coat.) This pattern is the most straightforward way that genes have of influencing what individuals are like.

The study by Anderson et al. provides evidence that breeding with the domestic dog brought the black gene into the wolf population. They also show that the increase in black wolves living in the boreal forest is a result of selection, meaning that the black color is an advantage for wolves in the forest.

The wild population of wolves is vulnerable because of environmental change and habitat loss. Variation from domesticated animals may not only endure in the wild, but may actually increase the ability of wild populations to survive. The idea that a trait such as dark coat color in wolves that is helpful for survival in a wild species originates in a close relative such as the domestic dog is an intriguing variation on our usual views of the interconnectedness of different animals.

**Gene flow between domestic dogs and wolves is meaningful from an evolutionary perspective, and it is likely that such gene flow occurs in genes related to traits other than coat color.**

## Teach Your Dog Not to Bark

***June 29, 2009***

It's common for dogs to bark out the window at people going by, and most dogs don't stop until the passersby are out of sight. This behavior is not confined to dogs who react to people in every context. Some dogs who are polite and friendly in social situations do it, too.

There are many reasons why dogs bark out the window. They might be barking out of frustration over not being able to greet the people. Dogs, like people, are not at their best when frustrated. Or, they may have learned that when they bark, the people go away, so they bark to make that happen. Alternatively, the barking may serve to alert the rest of the family to an intruder on or near the property. No matter what the reason for their barking, yelling is unlikely to help. When you yell, "Quiet!"

it sounds to your dog as though you are barking, too. Barking is contagious, so it's not surprising that when you join the frenzy, your dog continues to contribute to it.

One effective way to handle this problem is to teach your dog that when someone walks by your house, she should turn away from the window to get something that is valuable to her. To teach her to go away from the window when a person goes by, be ready with a toy such as a hollow rubber Kong stuffed with top quality smelly treats. As soon as your dog notices someone walking by, put the stuffed toy by your dog's nose, say, "This Way," and lure her to a place where she can no longer see out the window. Then, give her the stuffed toy so that she is reinforced for coming away from the window and she has something enticing to keep her occupied for a while.

Keep in mind the motto, "The treat is mightier than the voice," and remember to get your dog's attention via her nose rather than her ears. The treats have to be very high quality—skip the dry bones and kibble and instead use steak, chicken, cheese, or liver.

After a few repetitions in which she willingly leaves the window, start saying, "This Way," even before you put the Kong by her nose to lure her away. With practice, you should be able to say, "This Way," and have her come away from the window to get her treats without having to lure her.

Once your dog makes the connection between passersby and getting a toy stuffed with food, she will see people through the window and come running away from the window even without you having to say, "This Way." When she does this, give her a stuffed Kong to let her know that she did the right thing. Reward yourself, too, because you have successfully taught your dog a new and better response to seeing someone go by—turning away from the window rather than barking like a lunatic. Eventually, you can phase out the Kong by using it only sometimes, rather than every time she chooses the new behavior over barking out the window.

**With enough practice, the action of turning away from the window can occur in response to a new cue, which is the passersby, as well as to the verbal cue of "This Way."**

## Play and Pause, Play and Pause

***July 27, 2009***

Dogs love to play, and people love to watch them play. That is, unless the play seems out of control and likely to overwhelm, frighten, or injure one of the dogs involved. In that case, one or more dogs may be having a miserable experience, and watching such interactions can be aversive, even stressful.

It's tricky to differentiate boisterous, exuberant play from the kind of play that can lead to real trouble. Canine professionals make observations about many facets of the behavior to evaluate whether the play is appropriate or not. I think one of these, the presence of regular pauses in play, is particularly informative.

When play between two dogs has no pauses, trouble is much more likely to happen than during play with lots of breaks. Unlike human children who tend to play continuously with one another, dog play has a rhythmic cadence of full speed alternating with pauses. When I watch dogs playing together, whether it is rough-and-tumble wrestle play, pouncing play, or chasing play, I often say to myself, “Play, play, pause; play, play, pause,” as the energy of their play session ebbs and flows. The pauses need not be long—even half a second will do—but they should occur frequently.

Dogs can do different things during these brief pauses. Sometimes they just stop running, or stop wrestling and separate from each other. Some dogs look around and some lie down, but perhaps the most common action during a pause is a play bow, which is a signal that the dog wants to start playing or wants to continue playing. The play bow is the most frequent play signal in dogs, and involves dogs lowering the front end of the body and resting on their elbows, with their back end up in the air. Tails are often wagging, and dogs most often look at their play partner while performing a play bow.

When dogs are playing appropriately with one another, including lots of brief pauses in their fun, it often shows that both dogs are exhibiting emotional control. If they have the emotional control to execute frequent pauses, they most likely have the emotional control to prevent the play session from escalating into an out-of-control brawl or into aggression, either of which have the potential to result in serious injury.

Pauses allow so many good things to happen in a play session. There is the possibility for play signals, which lets the dogs reaffirm that all this leaping, jumping, and pouncing is playful in nature rather than being predatory or aggressive. Pauses let each dog decide whether to continue playing or perhaps to switch to sniffing the grass, chasing squirrels, or starting to play with someone else instead.

Play is often restarted with a chase, a play bow, pawing at another dog, or brief inhibited bites most often directed at the forelegs of the other dog. If both dogs resume playing after the pause, it is a good sign that they both want to be involved in the play session. That's important because the single most important rule of canine play is that everyone is a willing participant.

**Play signals, pauses, and the sense that everyone is having a good time are among the most important signs to check for when determining if dogs are playing nicely.**

# Does Your Dog Pee When She Greets You?

*August 10, 2009*

It's no fun when your dog greets you by the door, wagging her entire body, and spraying the floor with puppy pee as she wiggles about. It gets even less fun when this happens every evening for several months so that your coming home ritual always includes a few minutes on your hands and knees cleaning up the mess. What happened to the martini and recliner ritual of yesteryear? If you've got a dog who in all other respects seems to be house trained, but who greets you and other people by releasing at least part of the contents of her bladder, she is most likely performing what is called "submissive urination."

Submissive urination is a particular type of house soiling and it is unrelated to house training. A dog can be perfect at going outside to eliminate when the urge hits and still deposit urine in the house during greetings. It bears repeating that this is not a house training issue. It is actually a social issue and is not uncommon. Dogs who urinate when greeting people or other dogs often share several characteristics. They are typically under a year old, they are often very sweet and social, and they generally wag their tail and whole body while lowering their body toward the floor during greetings. The amount of urine that comes out is usually small and it often leaves a wavy pattern of lines and droplets along the floor rather than a puddle.

While this problem can certainly be annoying, there is good news for anyone who has a dog with this issue. First, most dogs outgrow this behavior by the time they are a year old, so at least it tends to be temporary. Second, dogs with this issue almost always have lovely temperaments, so while the urination can be irritating and a pain to clean up, the fact that dogs greet in this manner actually speaks well of them. Ironically, this urination during greetings is showing respect for the other dog or person, and a dog who is behaving respectfully is a dog who tends to be polite and biddable. Third, this problem is relatively easy to treat, and most dogs respond well and quickly to the treatment.

Because the urination occurs during the greeting, the treatment must occur in that context. Keep all greetings low key and do not directly approach the dog. As soon as the dog is visible, throw treats over her head so that she must turn away from you in order to find and eat the treats. Don't reach for or look at the dog until after she has consumed the treats. Preventing interaction until after she has received food from you often results in a "dry" greeting. To make this easier to live with until the dog no longer submissively urinates during greetings, consider greeting her outside or on hard floor surfaces whenever possible rather than on carpet so that cleanups are as easy as possible. Never punish her for submissive urination either physically or even with harsh words. In addition to being unkind, this will only worsen the problem and make treatment harder in the future.

**I continue to be called to consult about dogs who pee in greetings, and I continue to take delight in letting people know this behavior speaks well of their dogs even if it is a nuisance.**

## Keep Your Eyes on Dogs' Mouths

***September 14, 2009***

While the eyes may be the mirror to your dog's soul, I spend more time paying attention to the mouth. This is partly because I specialize in helping aggressive dogs, and these animals are armed with the equivalent of steak knives in their mouths.

Eyes may be intense, but no client of mine ever ended up at the hospital or in court because of what the dog's eyes did to them. On top of that compelling reason, the mouth of a dog can give a lot of information about the dog's emotional state and what he is likely to do.

One of the first things I notice about a dog is whether his mouth is open or closed. A relaxed, friendly dog will usually have an open mouth. The typical image of a happy dog is one with a relaxed face whose tongue is hanging a little out of his mouth as his lower jaw rests wide open.

An open mouth on dogs who are overheated tells me nothing about their emotional state because they will have their mouths open to cool down by panting no matter how they are feeling. On the other hand, a dog who is not warm but is nonetheless panting is probably feeling anxious. A dog who is fearful or nervous most often has his mouth tightly shut.

When dogs become emotionally aroused, their mouths reveal a lot about their internal states. Specifically, it is informative to observe the position of the lips and how visible their teeth are. When the corners of a dog's mouth, or commissures, are pulled back tautly, it is a sign that the dog is fearful, and this facial expression is actually called a "fear grimace." In a fear grimace, many of the dog's teeth are visible. This posture of the mouth connotes fear rather than the intention to attack. Fearful dogs may still behave aggressively out of panic, but many will take another way out of the situation if possible.

When an aroused dog pushes the commissures forward so that the mouth becomes O-shaped, the signal is a threat called an "offensive pucker." This signal indicates a dog who is willing to go on offense and who is most likely very confident.

Another visual signal involves retracting the lips to bare the front teeth. Showing the teeth in this manner is called a "tooth display." This signal translates best as "I have teeth, and I am willing to use them." In other words, a dog who is doing a tooth display is more likely to go on the offensive and behave aggressively than a dog who is not performing this display. Dogs who give a tooth display combined with an offensive pucker are especially likely to behave aggressively. However, many dogs do make threats with one or more of these signals without ever causing injuries.

Is the mouth open or closed? Is the dog panting? Does the dog display an offensive pucker, a fear grimace, or a tooth display? Answering these few questions can reveal a lot about how a dog is feeling and what he might do next.

**Reading dog visual signals is a critical part of understanding dogs, and of staying safe when those dogs may have the potential to act aggressively. The mouth is the source of potential injuries, but it is also the source of so much information.**

## The Dog Nose Knows!

***September 28, 2009***

Our dog, Bugsy (half black Lab, half handsome stranger), was not the cleverest of dogs, and we eventually learned to embrace the moments when he showed his lack of Einsteinian brain powers.

One day on our farm, he suddenly began to sniff his own footsteps in the snow, backtracking for several hundred feet along his own path. I chuckled to myself about his silliness, and even said aloud (to no effect), "Those are your own footsteps, you nut." Finally, he veered away from his own path and continued to track. Upon investigation, I realized that he was tracking a rabbit and that he had first caught the trail in his own paw print. He had stepped onto a rabbit paw print hundreds of feet before, and yet there was enough scent to get his attention and for him to follow. It made me wonder how many other times I erroneously identified his behavior as resulting from low IQ rather than existing in a different sensory universe.

It is a common mistake to assume that the perceptual, or sensory, world of humans mirrors those of all animal species, including dogs. Each animal species lives in its own such world, which in the behavior literature is called the Umwelt of the species. Their Umwelt determines what they tune into and what they do not attend to at all. While we humans are immersed in mainly visual and auditory stimuli, our dogs live in a world of smells. Scent plays a huge role in most aspects of dogs' lives, from tracking prey and hunting to competition between individuals, territorial marking, and mating.

Dogs have an organ in their nasal cavity that allows them to perceive molecules in the air. This organ, called Jacobson's organ, functions as a combination tasting and smelling structure. Dogs essentially taste the air with the Jacobson's organ. Sometimes dogs hold the mouth half open and the head pointed up—a posture known as the flehmen response—in order to better perceive the chemicals in their immediate environment.

The information that dogs receive from olfactory cues is astounding. Some estimates suggest that dogs are a million times more scent-sensitive than humans. Just from scent marks in urine and feces, for example, dogs can determine the sex, social status, point of the sexual cycle, and age of the individual who left the scent mark.

The tremendous abilities of dogs to detect even the faintest odors are a great help to humans. Besides tracking down lost people or cadavers in wilderness areas, underwater, or buried under an avalanche, they can sniff out drugs, explosives, weapons, land mines, and all manner of substances banned from importation including fruits, meats, vegetables, wildlife, or undeclared currency. They can also detect termites or bed bugs in houses, pirated DVDs, and in some cases, melanoma cells in human skin. There is evidence that dogs can use their sense of smell to predict epileptic seizures in humans, and to detect low blood sugar levels in people with diabetes.

Given that dogs' olfactory abilities are far superior to those of humans, I probably seemed clueless to my dog Bugsy as often as he seemed that way to me.

**It's fascinating to consider how ridiculous we must seem to dogs when it comes to our olfactory abilities. It's a reminder to be empathetic to others, no matter what species they are.**

## An Ode to "Wait"

### *October 5, 2009*

Years ago, a plumber came to our house to replace a broken water heater in the basement. I told my dog "Wait" at the top of the stairs to avoid a stampede as we went down. After the people had descended the stairs, I said, "Okay" and my dog came down to join us. The plumber commented, "That's a nice dog you've got. If only people could get their kids to listen as well!" The pride I felt in my dog for making such a good impression was a reward for all the effort I had put into his training.

Among all the cues that I have taught dogs to respond to over the years, "Wait" is my favorite. This cue tells your dog to pause and not to move forward. It is different than "Stay," which means to remain sitting or lying down in the same place until told that getting up is allowed. "Wait" prevents the commotion so typically associated with the start of a walk as dogs knock people over in their enthusiasm to get through the door. It's also a way to keep your dog safe from the tragedies that occur when dogs leap out of a car or a house and are hit by cars.

A dog who knows how to wait will respond to this word, even in the presence of a wide-open door, by pausing at the threshold and waiting for you to give her permission to proceed with a release word such as "Okay." The cue "Wait" results in a dog who is attentive to you even when excited about going somewhere, because she has learned that the way to get what she wants is to be patient and polite rather than rude and pushy. One of the best things about Wait is that it's surprisingly easy to teach.

To begin, go to the door you usually use when departing on a walk and put your dog on a leash. Stand with your back to the door so that you are between the door and your dog and your dog is a few feet from the door. Say, "Wait" in a calm, confident tone and open the door partway. If she tries to bolt past you and head outside, block her access with your body (not by using the leash!) so that she cannot get past you. If she pauses, step aside to give her a choice. If she chooses to Wait even though she has a clear path to the outside, say, "Okay," and let her through. If she chooses to make a break for it, block her access by stepping in front of her. Keep giving her a choice until she makes the right one (which is to wait for permission before going through the door), then say, "Okay," and let her go outside.

Teaching Wait to your dog will make departures peaceful rather than hectic, and is one of the easiest ways to present a polite, patient dog to the world.

**The lifesaving and convenience-inducing benefits of having a dog who responds to being asked to "Wait" continue to inspire me to teach this cue to dogs.**

## Holiday Pet Safety

***December 7, 2009***

Safety during the holidays is an issue for everyone: Drinking and driving is as unwise this time of year as at any other, candles on trees pose an extreme fire risk, and old fruitcake is a bad idea according to four out of five dentists who want their patients to have pain-free teeth over the holidays.

Safety issues also abound for animals during the holiday season. Amid the beauty of the festivities reside many perils for our pets.

Fire is a risk for pets, especially rambunctious young or even older athletic ones. Whether the candles in your house are kindled for 8 days, serve as a reminder of the light that appeared in the east, or simply provide ambiance, open flames require constant tending. Put all candles out of reach of pets and make sure neither the flames nor the pets are unattended.

If your decorations involve electricity rather than fire, be aware that power cords can look like chew toys to many animals. Keep them taped down or out of reach to lessen the chance of burns or shocks. Keeping holiday lights unplugged when you are not home is another way to help protect your pets.

Many holiday plants are poisonous to pets. Among the most dangerous are mistletoe and holly, especially the berries of these plants. Ivy and poinsettia are also dangerous to pets, but their toxicity levels are lower. Even the seemingly angelic lily is poisonous. Poisoning risks are not limited to plants. Chocolate can be deadly for both dogs and cats. Theobromine, the toxic chemical that causes the damage, occurs in especially high levels in baking chocolate—8 to 10 times the level of milk chocolate. Consuming chocolate makes many cats and dogs vomit or get diarrhea, while still others react with convulsions and seizures. Grapes and raisins can lead to kidney failure in dogs and are therefore potentially fatal to them. Cats are thought to be susceptible to this toxicity as well, but fruit-loving cats are rare. To be safe, keep grapes and raisins away from your pets as diligently as you keep chocolate away from them. If you even suspect that your pet has ingested any of these items, contact your veterinarian immediately to find out what to do.

Food doesn't have to be ingested to cause harm to your pets. For example, hot turkey and ham drippings smell so enticing that many animals leap onto the counter to get them, and in doing so, tip the pans and get scalded. Many holiday items, including food, pose choking hazards for dogs and cats. Among the most common threats are candy canes and other sweets, ornaments, small toys, and turkey bones, which splinter easily. Both ribbons tied around pets' necks and tinsel can strangle pets, and though this may be rare, it does happen to some dogs and cats every year.

Avoiding the potential hazards of the holidays allows the celebrations throughout the season to remain joyous for all the species in your home.

**Looking out for our pets is a year-round endeavor, but the holidays do pose numerous extra risks.**

# Teaching Your Dog to Stay

***January 4, 2010***

Having a dog who knows Stay makes life safer for your dog and easier for you. When visitors come to the door, a dog performing a Stay can make the situation pleasant for all rather than chaotic. I once had a glass bowl shatter in the kitchen, which could have been injurious to my dog's feet. I put him in a Stay in the living room, and then went to clean up the mess before releasing him. His rock-solid Stay turned what could have been a horrible situation into a manageable one.

Teaching a dog to stay is a methodical process. You teach your dog what you want him to do and reinforce him for doing it. You repeat this process many times, making the task gradually more difficult for your dog. The three "Ds" of difficulty are distance, duration, and distractions. Start with all three "Ds" at a low level—remain right next to your dog (low level distance) for a very short Stay (low level duration) when nothing else is going on (low level distractions).

The three most common mistakes when teaching Stay are:

1) Not releasing your dog from the Stay. Stay is a two-part behavior. "Stay" tells your dog to remain in place and not move from that spot until he is told that he can. A release word such as "Okay" or "Free" is the signal that gives him permission to get up. If you don't use the release word regularly, then most dogs become confused about how long to stay. Until you ask him to do

something else? Until you're not looking at him anymore? Until he gets bored? By consistently using the release, you teach your dog that he must stay until you tell him the Stay is over.

2) Giving your dog a treat after you release him from the Stay. Instead, give your dog treats while he is in a Stay so the Stay is the fun part. If he gets treats once he's released, then his emotional state during the Stay will be one of eager anticipation: "Oh boy, I'm going to get a treat soon, I can't wait, it'll be great, oh boy, oh boy OH BOY!" By giving him treats while he stays, he enjoys the Stay itself and won't be so excited and likely to get up before he's supposed to.

3) Making the Stays early on in training overly hard by asking for too much in terms of distance, duration, or distractions. It takes tons of practice (think hundreds of repetitions at gradually increasing levels of difficulty) to get a dog to be able to stay for a few minutes, when you are at some distance away from him, or there are distractions such as people, toys, other dogs, or squirrels in the area. Start easy and work up to more difficult Stays gradually.

The effort you put into teaching your dog Stay will be worth it. There will be countless occasions for you to reap the benefits of your hard work.

**I love teaching dogs Stay because I love having dogs around who know how to perform this behavior. Making life easier is always a goal, and Stay can help.**

# Visual Versus Vocal Cues

## *March 29, 2010*

Mimi was so pleased that in just one session she had trained her dog, Jake, to sit that she called her husband, Will, over to see their progress.

"Sit," she said to Jake, who wagged his tail and looked at her. "Sit!"—more wagging. "Sit! Sit! Sit!" Jake continued to stand there. After her husband walked away, Mimi said, "What's up, Jake?" pointed her finger at him and said, "Sit" at which point Jake did. What's going on here?

The answer is probably that Jake has learned how to sit in response to a cue from Mimi, but that cue isn't the word "Sit." Jake is probably responding to the pointing motion of her finger and not to what she is saying, and when Will was there, Mimi didn't point.

Humans most often express themselves vocally, whereas our dogs communicate primarily with visual signals. Dogs are watching us, and we are talking to them. This explains so much of the confusion between our otherwise largely compatible species. Dogs can't figure out what their humans are trying to convey, and we can't figure out why our dogs aren't listening.

Dogs often pick up on visual cues that we use, inadvertently or not, when training them. So, if, during training, we use a hand gesture while saying, "Sit," most dogs will learn that the hand gesture means to put their bottom on the ground long before they figure out that the word "Sit" means to do the same thing. If we are aware that the visual signal overshadows the vocal one, problems can be avoided. We can teach our dogs to sit when both cues are given simultaneously and then actively teach the vocal cue once the visual one already means something to our dogs.

The key is the timing. Begin by teaching your dog to sit using both the vocal and the visual cues you want your dog to know. Once he responds to them together, start separating them. Give the verbal cue "Sit" a half a second before you give the visual cue, such as a pointed finger. After many repetitions of sitting and getting reinforced for doing so, your dog will learn that the word "Sit" predicts the hand gesture whose meaning he already knows. He will eventually respond by sitting when he hears "Sit" even without the visual signal.

Research has shown that dogs learn visual signals faster than vocal signals. Therefore, it is most likely that if your dog is sitting when presented with both cues, he already knows the visual cue on its own. To check for sure, you can experiment by giving just the visual cue and see if your dog sits.

Many people, including Mimi, have inadvertently done the reverse experiment in which they tested their dog's ability to respond to the vocal cue alone, and found that he does not. Showing off what our dogs know seems to make many of us inhibited about using a lot of motion, and we choose just to speak instead. The result is that a dog we think knows how to sit in response to a verbal cue appears to have no idea what is being asked of him.

**The tendency to attend to visual cues over vocal cues tops my list of things I wish everyone knew about dogs.**

## Play Should be Fun, Not Tense

***May 3, 2010***

I am often asked, "How can I tell if my dogs are just playing or if something is wrong?" It's a tough question to answer without watching the dogs myself, but some basic guidelines are useful in evaluating canine play.

Dogs who are playing appropriately have a lively, carefree quality. Proper play usually involves chasing, parallel running, rolling around together, batting with paws, and brief pounces at one another. Play is often very bouncy in nature with many behaviors of dogs being exaggerated in the vertical direction. In appropriate play, there are frequent play bows or other pauses in the action. Dogs' mouths are usually open, and all dogs are willingly engaged in the activity.

Vocalizations tend to be constant in pitch rather than suddenly deepening or becoming high pitched. Usually, all the dogs spend some time with their head down low, either while play bowing, rolling on the ground, or simply as they approach another dog.

Frequently, dogs are holding back a bit. That is, if they are running, they are not going at top speed every second, but may be going a bit slowly sometimes to encourage another dog to give chase or to catch up. If they are wrestling or pouncing, dogs are sometimes using only a fraction of their strength. When they are mouthing at one another, they are never biting down with the full force of their teeth and jaws.

Perhaps most important, dogs who are playing seem to be having fun, rather than engaging in a serious activity. Judging this is a bit subjective, but still very informative. Generally, this fun quality is incompatible with the intensity of more serious types of behavior.

Signs of tension may mean that dogs are starting to take things seriously, which can lead to trouble. If any dogs seem stressed, tense, or uncomfortable, then the mood may no longer be playful.

Nervous dogs often stick their tongues out in flicking motions. Sometimes the tongue just peeks out between the lips in a little darting motion. Other times, the tongue goes out quickly and arches way up and over until it touches the nose.

Other signs that a dog is scared, nervous or otherwise uncomfortable include trembling, excessive salivating, whimpering, attempting to escape or hide, cowering, whining, yelping, and panting that is not heat-related.

Tension in the form of stiffness throughout the body is a bad sign. Dogs who go stiff are not playful, and in fact, may be about to bite. The instant that any dog in a play group goes stiff, it's time to change the mood and get the dogs moving again, preferably away from one another. Going stiff is not to be confused with pausing in play. A dog can stop and remain in one position, but doing so as a pause in play with a relaxed body is a good thing, whereas going stiff with tension in the body is a sign of potential trouble.

If in doubt, interrupt the play bout and give the dogs a break before letting them continue. If it's not fun for all the dogs involved, it's not appropriate play.

**There's an art to evaluating play that only comes from experience, but general guidelines are a good place to start.**

## Dogs at the Farmers Market

***June 21, 2010***

I love it that dogs are welcome in so many places in Flagstaff, including the Farmers Market on Sunday mornings, where the presence of dogs makes the whole experience better for most people. Yet, I sometimes see tense situations involving dogs there, and I've even occasionally seen some harsh words exchanged between people because of dogs.

That's a real shame because I really think the sight of people out enjoying life with their dogs dresses up a city considerably. By following a few simple guidelines, people with and without dogs can avoid these skirmishes.

1. The Farmers Market is not for every dog. Some dogs love the crowds, the smells, the kids, the other dogs, and the live music. Other dogs don't. If your dogs are not relaxed, social, and friendly with people of all

ages and other dogs, all in a chaotic situation, let them stay home.

2. If you want to pet a particular dog, ask before you approach the dog. If the owner says no, respect that.

3. Don't let your dog lick people. Dogs are especially likely to lick children because they are low down and sometimes wearing food.

4. Teach your dog not to jump up on people. Not all people love dogs, and even those of us who adore them don't always appreciate being jumped on. Elderly people or kids can easily be knocked over, and many people have medical conditions that make it a serious situation. People with insulin pumps, inner ear disorders, and various skin conditions can be injured by jumping dogs. Once when I was badly sunburned, a dog jumped up on my red legs with sharp nails. When I said, "Ouch!" the owner looked at the marks left on my legs and said, "Oh, she scratches me all the time like that. It doesn't hurt." I found that rude and annoying. Go figure.

5. Know the signs of stress in dogs and remove your dog from any situation that is creating such anxiety. Tucked tails, shaking, cowering, excessive panting that is not heat related, tongue-flicking, actively avoiding looking at another dog, trying to hide behind you, or trying to get away are all behaviors that indicate the dog is uncomfortable with the situation.

6. Don't let dogs greet each other unless both owners have agreed to it. It's not fair to assume that every dog is comfortable having other dogs sniff or touch them. If they are allowed to greet, pay attention to them as they

meet to avoid letting the social encounter go bad or allowing leashes to become entangled. Maintain slack in the leashes to avoid passing tension down those leashes to the dogs, and be aware of signs of stress in either dog.

Having dogs at the Farmers Market is my kids' favorite part of the whole event, and in many ways, it's mine, too. Following these basic guidelines will help dogs to continue to be welcome and happy there.

**Sadly, my advice was not heeded by enough people (or even seen by enough people, to be honest) and not long after I wrote this, dogs ceased to be allowed at this weekly Sunday event, which is actually called the Flagstaff Community Market.**

## Cutting the Nails of an Unwilling Dog

***October 4, 2010***

Cutting a dog's nails is rarely fun for anyone. Many dogs are afraid and struggle madly, which makes it even harder for people to avoid inflicting a small but painful injury.

I spent almost a year working as a dog groomer, so I have cut the nails of dogs ranging from docile to devilish. No matter how they behaved, each one had to go home with properly trimmed nails, so I know a few tricks about getting the job done.

One of the easiest ways to cut a dog's nails is to keep her occupied while you do it. A Kong® or other hollow toy stuffed with something delicious such as steak, hamburger, peanut butter, turkey, or cream cheese can often help your dog to relax and ignore what you are doing. A real bone, a chew toy, or any special treat will work. Let her have it long enough to get really involved with it before you bring out your clippers. Give her this same treat at other times so that she doesn't get anxious and think, "Uh-oh, here come the clippers!" whenever the Kong® appears. Another option is to clip just one nail, and then immediately take your dog on a walk. When I say, "immediately," I mean that you cut one nail, and the very next second you grab the leash, and say, "Let's go for a walk." After doing the one-nail-a-day-and-go-for-a-walk sequence for several weeks, many dogs will start to be happy when they see the nail clippers because they have learned to associate having a nail clipped with going for a walk. Once your dog is happy about having her nails clipped, you can do more than one at a time, working up to a whole foot at a time, and then two feet, until finally, you can do all her nails at once before going on that walk.

Many otherwise lovely dogs behave aggressively when having their nails cut, and this does not put a dark blot on their character. Still, it's wise to be wary. If your dog bites or tries to bite when you cut her nails, it is perfectly acceptable to muzzle her. It's more humane to use a muzzle and trim her nails quickly than to struggle for a long time while perhaps not doing as good a job on the nails.

It's an expense and a hassle to take your dog to a professional such as a groomer, veterinary technician, or a veterinarian for nail trimming, but it may be worth it. As a behaviorist, I focus a lot on the relationship between people and dogs, and your trusting relationship can be ruined (or at least experience a setback) if your dog is frightened of you when you cut her nails. It's better for her to have those negative feelings about a groomer or somebody at the vet's office than about you.

All too often, nail clipping includes a tantrum or a wrestling match (and the dogs aren't always angels either!) I hope at least one of these ideas makes the process faster, easier, and less stressful for you and for your dog.

**During the pandemic, many more people cut their own dog's nails, making use of one or more of these techniques to get the job done.**

## Mixed Signals

### *January 17, 2011*

Here's a cheerful thought: Frustration plays a part in nearly every relationship.

For example, dogs who jump up incessantly are a big source of frustration for many people. Most attempts to change this behavior are doomed to failure because many things that we humans do are misinterpreted by our dogs, which leads to misunderstandings and confusion on both sides.

The natural response of many humans when a dog jumps up on them is to use their hands to push the dog down. This often results in a dog jumping even more exuberantly onto the person, who pushes the dog away still more firmly, which in turn leads to a dog leaping even more forcefully until finally the person shouts at the dog, who looks stricken or sad, and perhaps even cowers. What's going on here?

The person may legitimately be thinking that the dog should get the message to knock it off, as in, "What part of 'I'm shoving you away because I don't want you to jump up on me' don't you understand?" And the answer is "All of it."

The dog doesn't understand what the human is trying to communicate ("Cut it out and get off of me!") because the shoving that the human is doing means something else entirely to dogs.

Specifically, pushing at another individual with your arms (forelegs to dogs) is a play signal in canine society. That means it's an invitation to play. So, it's easy to see how a jumping dog who is shoved off of a person gets more revved up. When the message seems to be, "Let's play!" it's hardly surprising that the dog continues to leap up joyfully to get the game going rather than calming down and leaving the person alone.

Imagine for a moment that you actually said to the dog in English, "Stop jumping on me!" but those words literally meant, "Let's play!" in dog language. Since dogs' communication is based much more on what we do than what we say, that's a pretty good description of what's going on. No wonder it's frustrating for both people and dogs. The person can't get the dog to stop jumping up and the dog should hardly be blamed for thinking, in some canine kind of way, "Sheesh! Humans! They act like they want to play, but then when you try to play, they get mad about it."

Best friends we may be, but the fact that we belong to different species makes for a lot of misunderstandings between us. That can leave us feeling that our dogs are stubborn, or even not so bright. The next time your dog doesn't seem to understand you, pause to consider the possibility that your dog is thinking that YOU are the one who is either obstinate or not so clever.

To read more about differences in the natural behavior of humans and canines and how they influence the relationship between the two species, check out Patricia B. McConnell PhD's book *The Other End of the Leash: Why We Do What We Do Around Dogs*.

**Dogs and people understand each other better than most species seem to, perhaps because of the many thousands of years of history we share, and perhaps because of the time we as individuals have spent interacting with members of the other species. The earlier column from March 30, 2009, "Is Your Dog Like the Flintstones' Dino," offers practical advice for people whose dogs jump up on visitors.**

# Five Causes of Canine Aggression

### *March 7, 2011*

Does your dog growl at your children but behave like a model citizen around you? Does he bark and lunge only at men? Only at women? Does she try to bite anyone who reaches to pet her? Does she get snarly in the evening? The contexts in which a dog behaves aggressively create a pattern, and this pattern provides insight into the cause of that particular dog's aggression. Although there are many causes of it, most canine aggression can be classified into one of five major types. More than one factor may make a dog behave aggressively, so many dogs show the characteristics of more than one type.

**Fear**

The most common cause of aggression is fear. Dogs who are fearful often react to strangers but not familiar people or they react to men but not women. They typically react to large, deep-voiced men with beards or to people who are carrying things or wearing a hat. Many fearful dogs have a family history of fearfulness or were poorly socialized as puppies. If dogs react most vigorously when unfamiliar people approach or reach down to pet them, they are probably fearful.

**Status**

Although it's not actually particularly common, there's a tendency to assume that all aggression results from status-related issues. Dogs who have this type of aggression are more likely to behave aggressively to people within the household than to outsiders. They are also more likely to be aggressive to women and to children, especially the youngest member of the household. They may act aggressively when they are up on furniture or when they are approached while resting or sleeping.

### Arousal

Excessive stimulation and excitement can lead to arousal-based aggression, especially in dogs who have difficulty calming themselves down. These dogs tend to spiral out of control with increasing amounts of leaping, spinning, and vocalizing while they are aggressive. The aggression can be directed toward anybody in the immediate area, regardless of what the person is doing.

### Frustration

Dogs who act aggressively out of frustration are often prone to having fits when they don't get what they want. They generally have poor impulse control, no self-control, and little patience. The aggression is sometimes redirected to an innocent bystander when the dog sees something that he cannot get to because of a leash, a fence, or a window. Aggression may result when these dogs see something that they want but are prevented from having, such as a bone, a toy, or a play buddy.

### Pain

If aggression that is uncharacteristic of a dog's personality comes on suddenly in a fully mature dog, it may be because the dog is in pain. Another clue that the dog's aggression is a result of pain is that the dog has good days and bad days. Some dogs who are in pain tend to behave aggressively only in the evening or only after having had more exercise than usual. Others react only when a certain part of their body is touched.

If you have an aggressive dog, job one is to figure out the cause of the behavior. Only then is it possible to develop a plan to deal with the problem.

**Dogs can be aggressive because of more than one of these factors, but the pattern of their aggressive behavior is still the key to the cause of the aggression.**

## Preventing Canine Aggression Over Food

***May 16, 2011***

"I've been putting my hand in his food while he's eating since he was a puppy, so he's never growled at me over his food."

This sort of comment sets my teeth on edge because repeatedly bothering a dog who is eating is actually an effective technique for teaching dogs to behave aggressively around food, *not* a great way to prevent it. Many such dogs start to growl, snap, or bite when someone comes near their food. It's like they're saying, "Enough already. Leave me alone!" If a dog is constantly bothered while eating but never displays food bowl aggression, it shows that he's a great dog, not that harassing him was a good idea.

The natural response of many dogs when you approach, reach for, or take away their food is some canine version of, "Hey! It's mine! Back off!" Creating a response that's the canine equivalent of, "Oh boy, oh boy, oh boy, here she comes!" is a great way to prevent dogs from developing food bowl aggression.

You want your dog to feel happy when you approach him while he's eating, and even when you reach toward his bowl or take it away. Dogs who are happy about your approach are not going to growl or snap to get you to leave.

If you regularly walk by a dog who is eating and toss a treat to him, you are teaching him to anticipate a treat whenever you approach him at his food bowl. Once he learns that your approach predicts something good, he'll be happy to see you coming.

To begin, walk by your dog as he eats and toss a treat without stopping. Do this only 1–2 times during any feeding session and don't do it every time your dog is eating. Overdoing it can cause a dog to feel irritable, the same way many people feel in a restaurant when a server refills the water glass after every sip.

If your dog begins to look up in anticipation when you approach, he is ready for the next step, which is to walk toward him, stop, toss the treat, and then walk away. The step after that is to reach toward the bowl, toss a treat, and then walk away, and the last step is to pick up the bowl, put a few extra treats into it, and then give it back to your dog before walking away. It usually takes a few days to several weeks to work through each successive step.

This technique can prevent food bowl aggression. If your dog is already behaving aggressively around his food, or if at any point in this process your dog shows signs of aggression or tension (such as stiffening, growling, eating faster, hovering over the bowl, snapping, or showing his teeth), stop and seek help from a qualified trainer or behaviorist.

The result of this process is a sentiment that's a joy for me to hear: "My dog doesn't growl over his food because I taught him to love it when I come near him while he's eating!"

**Teaching dogs not to be aggressive around food is a great thing to do, but it's not very commonly done.**

## Grieving a Dog's Death

***June 6, 2011***

My brother-in-law, sister-in-law, and two nieces said goodbye to their dog today. Lizzie was almost 14 years old and in heart failure. They thought they would lose her last October, but she survived for months more than they expected. That doesn't make her death any less sad. Though it's great when a dog reaches old age, it doesn't take away from the fact that the lives of dogs are far too short relative to ours.

There are many ways to prepare for and handle the pain of losing a dog, all of which honor the dog's life and the relationship that you shared.

If you know that the end is near for your dog, and she is able, take her to her favorite place, whether it's the park or a place where she can swim. Treat her to her favorite foods and give her special items to chew on, so you can make her happy and provide yourself with fond memories of those last days. Take photos of these moments so that you can look back and know that the end of your dog's life was filled with kindness.

Tangible reminders of your dog can be wonderfully therapeutic after your dog has passed away. Photos are helpful for many people, especially if you have some good ones of your dog over the years doing the things you'd most like to remember—running through the woods, looking through the window as you come home, playing with a favorite toy. Even one of the many times she got into the trash can be a treasured memory after she is gone. A photo of your dog misbehaving in this way can be fun to look at, even if it was never fun cleaning up the mess. Many people like to make collages or memory books of their dogs with photos from all stages of life.

Other ways to create memories of your dog include putting together a record of your favorite stories about her. Write down what your dog most loved to do, how she changed over the years, the biggest trouble that she ever got into, and some of the happiest times you shared. Some people like to save a little lock of fur.

It can be healing to make a donation in memory of your dog to a shelter, rescue group, or any other organization that works on behalf of pets, as a way to pass on the joy that your dog gave to you.

Though it's challenging, refuse to allow anyone to give you the "it was just a dog" treatment or try to trivialize the pain you are feeling. It hurts to lose a family member no matter what the species, and you need time to grieve, even if not everyone understands what a big loss you have suffered.

There's no way out of the pain when a dog dies but to move forward through it. My hope for anyone who loses a dog is that over time, the sadness fades and the happy memories linger.

**In the more than 25 years since I have been working professionally with dogs, I have seen a societal change about understanding the pain of losing a dog. It used to be common for people to say to those who were grieving, "It's just a dog!" but now it is the norm for people to understand the pain of losing a dog and to be supportive of those going through it.**

# A Fearful Biting Dog

***July 18, 2011***

"Does he bite?" I asked.

"Yes," she answered.

"Does he bite children?" was my next question.

"Well," she said, "We don't want him to. That's why we brought him here—to get socialized."

I had approached this little dog to distract him when I noticed him in a staring contest with two big dogs. He was acting tough but was clearly terrified. My concern was that someone would run between them and be frightened or injured if the tension between the dogs escalated. The owner of the little dog immediately came over to tell me to stay away from the dog, which is when we had our conversation.

The mom in me was furious that she had brought a dog who bites to the park when lots of kids were playing their Sunday soccer games and tied him to the back of a goal. However, the canine behaviorist in me looked at the situation differently.

I know from over a dozen years of working with dogs with serious behavior problems (and specializing in aggression) that people often think they are doing the right thing with an aggressive dog even when they're not.

Although it is not widely known, most dogs who are behaving aggressively toward people are acting out of fear. The aggressive behavior diminishes only when the fear does.

Taking a fearful dog to a place full of people will not help him conquer his fear. It will actually make the fear, and thus his behavior, even worse because he is having yet another experience of being terrified. This is not intuitive because it's natural to think that if a dog is misbehaving around people that he should go to classes, to the park, or any other place with lots of people for socialization. Though this seems sensible, it's not actually true.

Socialization refers specifically to the process that occurs during the sensitive period—between 3 and 12–14 weeks of age—when puppies are becoming aware of the social world and learning how to behave within it. Anything or anyone that a puppy experiences in a positive way during this critical period of development is unlikely to produce fear in the dog later in life.

Proper socialization includes providing puppies many good experiences with new people during this brief and specific period of development so that they accept new people throughout their lives without being afraid. Only puppies can be socialized—not adult dogs.

Taking a fearful dog out and about to be with or near people is unlikely to help a fearful dog become less afraid. The exposure elicits fear in the dog, giving him additional experiences that confirm how scary it is to be around people. This usually makes the fear worse, along with any undesirable behavior that is a result of the fear.

Most people dealing with an aggressive dog desperately want to improve their dog's behavior, and knowing what *not* to do is an essential part of success. Though it may seem like a good idea, it's counterproductive to take fearful dogs who are behaving aggressively to crowded places.

**There is a lot of judgment from dog professionals toward people who do what the woman in this column does, which is unfortunate. People are trying to do right by their dogs, and we need to focus on educating people, not blaming and shaming them.**

## A Fetching Dog

***March 5, 2012***

The game of fetch wins the prize as the ultimate good-for-us, good-for-them activity. It gives dogs exercise without humans having to work up a sweat and is a great way to teach dogs the crucial skill of dropping an object on cue. Since it's fun and interactive, it enhances the relationship between people and dogs. Fetch provides the basis for some of the most amusing tricks, such as teaching a dog to go get a tissue when he hears someone sneeze or to grab a beer from the cooler when told that someone is thirsty.

Unfortunately, many owners expect their dogs to play fetch without any training, and they are disappointed when that doesn't happen. They assume that their dog just isn't into fetch. That's a shame, because many dogs who aren't naturals at the game love to fetch once someone has taught them how to play.

When teaching a dog to fetch, do what you can to get off on the right paw by choosing your dog's favorite toy, whether it's a ball, a plush squirrel, or something squeaky. Begin inside the house where the distractions are fewer and less intense than outside and it will be easier to keep your dog's attention. In the early stages, the farther you throw the ball, the less likely your dog is to go after it, so start with short tosses of 10 feet or less. Use two or three toys so that you can throw one when your dog comes back to you even if he doesn't want to let go of the one in his mouth.

After throwing the ball, either make rapidly repeated high-pitched noises such as woop-woop-woop (pup-pup-pup is also good) or clap as you run away from your dog to encourage him to run toward you. Change the throwing direction with each toss to keep it unpredictable. It's the unpredictability that makes much of play so fun.

Another way to keep it fun, so your dog will want to keep playing, is to throw the ball the microsecond he comes back to you. It's more natural to stand there holding it (your dog thinks you're hoarding it) as you praise him, but that may bore your dog and cause him to lose interest. Finally, quit before your dog wants to. "Leave 'em wanting more" is a great strategy when teaching fetch.

As wonderful as fetch can be, it's not the right game for every dog. Dogs who become aggressive either because they become too aroused or because they are possessive (defending their toys with threatening or even injurious behavior) may not be able to play fetch without expert help.

Any dog who is physically capable of playing the game can learn to fetch, and many dogs from a huge variety of breeds love it. It is true, however, that fetch is often picked up fastest by dogs who have a natural tendency to chase after things or are toy-motivated.

If you teach your dog to play this wonderful game, you will have a dog who is fetching in every sense of the word.

**If you want a dog who plays fetch, one important route to success is choosing a dog who is a natural retriever. If that doesn't work out, you can train your dog to fetch but that doesn't guarantee that such play will be done with great enthusiasm.**

## My Dog's Mystery Adventure

***September 17, 2012***

Aunt Liz and I were surprised when we pulled into her driveway to see my dog out in the yard because we had left him inside the house. Yet, there he was, lying in the shade by the back door when we returned after being out all morning. "Well," I said to her, "at least he stayed here. That's lucky." I was amazed because I would have bet good money that if my dog ever got out of a house with no fence, he would have taken his opportunity at freedom and done something with it.

It turns out this would have been a good bet, as we later learned. That afternoon, my aunt hosted a barbecue with many of her neighbors, and they all seemed to recognize my dog. He had met them earlier in the day during the course of his travels.

One woman told me he had come to her porch and he looked so hot and tired that she gave him a bowl of water, and then another when he downed that one quickly. "He sure was thirsty!" she told me.

It's pretty obvious that he had had quite a lot of fun working up that thirst earlier on in his wanderings. According to one neighbor, he had spent some time playing with her dog. "My dog isn't always so playful, but she just adored your dog. They had such a good time together. I was sorry to see him take off, but I'm glad to know who he is now. Can they play again the next time he's in town?"

Another person told me that he had seen him running along the Connecticut River while he was in his canoe. "That dog looked like he wanted to join me. He followed along until I hauled out, and then he stepped right into my boat. I love a dog who wants to be on the water." My dog did indeed love canoeing with us, but this is the first time he solicited a ride from someone else.

Many miscellaneous comments helped fill in the details about his activities: "That's the dog I saw chasing a squirrel in the woods." "He sat down by the stroller and my son was so happy!" "He is fast! I saw him running alongside the road and he was nearly keeping up with the cars." "When we saw him, he was picking blackberries along the trail. I've never seen a dog do that before!"

The more I heard, the more alarmed I became. I'm lucky that my dog wasn't hurt, but he could have been, especially since he crossed so many roads. I guess I'm also lucky that I learned so much about his escapades that day. Most people whose dogs have roamed around unexpectedly would give anything to know the details.

Thanks to the people of Brattleboro, Vermont, I have a pretty good idea about much of what happened during my dog's unplanned, unchaperoned, spontaneous adventure.

**I am well aware of how terrible it can be for dogs to escape, both because of the potential hazards they might face as well as for the safety and comfort of others in the area. I do not mean to make light of my huge goof in accidentally allowing it to happen. Yet, when I think back on Bugsy's life, I am thrilled to know that he had this glorious day filled with joy, and that he brought some joy to others as well. I am grateful to my aunt's dog Nellie (who really did hang out in the yard the whole time) for opening the door for Bugsy, who never acquired door-opening skills.**

# Contagious Yawning in Dogs

*April 2, 2013*

Long before videos and pictures could go viral, yawning could spread through a social group almost immediately. It is extremely fortunate that most diseases are not as contagious, because it's a frightening thought to imagine a world in which the flu or malaria could spread and manifest symptoms as quickly as yawns.

It has been known for many years that yawns can be "caught" in various species of primates, including chimpanzees, bonobos, humans, macaques, and baboons. In recent years, the contagious nature of yawns between dogs and people has received a lot of attention from scientists.

Multiple studies have shown that dogs can "catch" yawns from people. Dogs who observe people yawning are more likely to yawn than dogs who observe people opening their mouths wide without actually yawning. A recent investigation reveals that human yawns are contagious to dogs by auditory means. Translation: Dogs will yawn in response to hearing a human yawn.

In this study of auditory yawn contagion, dogs were more likely to catch familiar yawns than unfamiliar yawns. In other words, yawning was more contagious when the dogs knew the person who was yawning, which shows the relevance of a social variable to the contagion level of yawning between people and dogs. This is consistent with the finding that the emotional closeness between two individual baboons makes their yawns more contagious to one another. If we're anything like baboons (and I strongly suspect that we are), we should consider it a compliment when friends and family members or associates at business meetings yawn back at us.

Contagious yawning is of particular interest to scientists who study empathy, which is the capacity to recognize the emotions being experienced by another individual. In primates, yawn contagion is an indication that the observer has empathy for the yawner. It was thought that in dogs, yawns were induced by a hardwired behavioral pattern that was simply exhibited in response to a releasing stimulus. However, the observation that dogs yawned more in response to the sound of familiar yawns than to the sounds of unfamiliar yawns suggests that empathy may play a role in the contagion in dogs, just as it does in primates.

According to one study, puppies under the age of 7 months are not susceptible to yawning contagion, showing a developmental aspect to the behavior. This may indicate that dogs don't acquire the ability to empathize with humans until they reach this age.

It's easy to overlook how fascinating it is that this social contagion occurs between two different species. We're so used to considering dogs our family members and best friends that we can lose perspective on how astounding that is. It's worthwhile to ponder the biological wonder of the relationship between dogs and people because it's unusual to have such a close and long-standing association across species.

Reading and writing about this subject has given me a serious case of the yawns, along with the dog sitting right next to me. We're obviously both social, empathetic individuals with a close relationship!

**Since studies first came out about this phenomenon, I have spent a ridiculous amount of time yawning at dogs I'm spending time with. It is really gratifying to watch them yawn in response because it confirms my strong feelings of connection to them. On another note, I wrote this column years before COVID-19 and it's odd to read my fears of contagion before the pandemic.**

## No Runner's High for Ferrets

***May 7, 2013***

If people go running by your home accompanied by their cats, geckos, ferrets, parrots, or rats, then you live in a very different neighborhood, perhaps even a different world, than I do. It's just not part of our relationship with those species, however close we may be to them. Yet running is something that many of us share with our dogs. Dogs typically like running, which is yet another commonality between dogs and humans.

Okay, perhaps not *all* humans. In recent years, people have become more sedentary than at any time in our history, and many of our dogs have joined us on the couch. As a result, that rush of good feelings we get from running—the runner's high—is no longer as widespread. Yet, the potential to activate the chemical reactions that cause the runner's high still exists in both dogs and people.

The runner's high is caused by neurotransmitters called endocannabinoids that signal the reward centers of our brains. They lessen pain and anxiety as well as create feelings of pleasure and well-being. Running results in higher levels of endocannabinoids.

Researchers at the University of Arizona investigated the production of endocannabinoids in multiple species as a result of running. These scientists wanted to know whether species that do a lot of long-distance running evolved to like running via the pathway of reward centers in the brain. The researchers predicted that running would result in chemical changes that are associated with pleasure in species with a history of endurance running, but not in species whose natural history does not include such running.

They compared the effects of running on endocannabinoid levels in three species. Two of the species—dogs and humans—come from long lines of endurance runners. The other species, a ferret called the European polecat, is not an endurance running species, though they can run quickly over short distances.

The researchers trained dogs, people, and ferrets to run and walk on a treadmill and took blood samples from them before and after each session. Blood taken from dogs and people after running contained highly elevated levels of one endocannabinoid called anandamide. Blood from ferrets after running did not have elevated levels of any endocannabinoids. None of the species had elevated levels of any endocannabinoid after walking.

Dogs and humans receive a pleasurable chemical reward for running, but this quirk of brain chemistry that makes both dogs and humans love running is not universal among mammals. Ferrets, for example, derive no such benefit. The scientists who conducted the study concluded that chemical changes from running and subsequent effects on the brain help endurance-running species enjoy running. The brains of dogs and humans are primed to like endurance running, which may have provided the evolutionary mechanism necessary for us to develop such skill at it.

This study shows that dogs and humans, unlike ferrets, achieve the runner's high. A friend of mine who hates running believes the study also suggests that she is part ferret, but scientific evidence to support this claim remains elusive.

**The friend mentioned in this column is the well-known canine behaviorist and dog trainer Patricia McConnell, PhD, who loves hiking, dogs, and ferrets—but not running.**

# Puppies Can Make You Dog-Tired

*June 18, 2013*

Ahh, puppies. Though they are adorable, fun, and lovable, raising a puppy has challenges that are as much a part of the process as all the tender moments.

I once actually said to my husband in despair, "Whose idea was it to get this puppy, anyway?" It had been mine, but I was overlooking that detail in my desperation to find someone to blame for our situation.

We were fostering a puppy and were therefore sleep-deprived, overwhelmed, and groggy. Our puppy was an 18-week-old, field-bred English Springer Spaniel puppy who loved people and got along with all dogs, including our adult male. She came when called, would sit and lie down on cue, and stayed with us on walks all over our 150-acre farm.

Unfortunately, she wasn't house trained, wouldn't eliminate while on a leash, and worse, she would eliminate in her crate. Once, she defecated in my office while I was on the phone with a client discussing house training. As it would have been unprofessional to interrupt my client, all I could do was recognize that this was a setback and accept the ironic absurdity.

Also on the downside, she had enormous exercise needs. She ran for hours on the farm with us multiple times daily, which tired her out enough to relax for about 15 minutes afterward. She used her sharp puppy teeth on us, as puppies tend to do, and didn't respond to the standard techniques for lessening such mouthy behavior. She was endearing and we loved her, but she was driving us crazy.

Caring for a puppy is a lot of work, even for a professional, and if you ever forget that, puppies will remind you in a way that is demoralizing, humbling, and possibly ridiculous.

It's enough to make anyone respect female dogs, who take care of not just one or two puppies, but seven, eight, nine, or more. Females perform all parental care in dogs, including nursing the puppies, protecting them from dangers, and retrieving them if they wander away. They even lick their puppies to stimulate elimination.

Though many people think that all fathers will hurt their puppies and must be kept away from them, this is not universally true.

The general wisdom is that the only dog who should be around the puppies for the first 2 or 3 weeks is the mother. This prevents injuries to the puppies and protects other dogs from the mother. Later on, visiting rights depend on the father's personality. If he is a gentle dog who is comfortable around dogs of all sizes and ages, then he will probably be fine around his puppies.

Even the most benevolent father needs to be supervised constantly when he is with the litter and removed immediately if he is too rough or seems uncomfortable.

The female dog can take care of them by herself anyway, which is impressive. Though I love puppies, even *sharing* the responsibility of raising a single one, no matter how delightful, exhausts me.

**Every time anyone shares with me that they have a new puppy, I am filled with a mixture of envy (Aww, a puppy!) and sympathy (Oh, a puppy!). I often offer to give them a break by watching the puppy for a bit, which does both them and me a lot of good!**

## Body Blocks

### *August 6, 2013*

An influential dog controls other dogs' movements by managing the space around them. She puts her body in front of another dog, blocking the way that she does not want him to go, and leaving access to the way that she does want him to go. If she uses her body to control the space but the dog keeps coming, a confident dog will hold her ground. She doesn't collide with the dog, but she is committed to controlling that space.

These space management maneuvers can be adapted from dogs to humans and become part of anyone's training techniques, though they are not recommended for use with aggressive dogs. With adjustments to allow for human anatomy (especially that two-legged thing), you can use body language that dogs understand to communicate what you want them to do in many contexts. Such moves, collectively called "body blocks," all influence dogs' behavior by managing the space around them. Using body blocks prevents having dogs who only respond when both the human and the dog are attached to a leash.

The effective use of body blocks requires attending to your body position and movements. Imagine that a dog who has been told to stay starts to get up and investigate a treat that dropped on the floor. If he moves forward toward you and to one side, you can counter him with your own motion, stepping forward and sideways with one step into the space that he was about to occupy. If he pauses, then you should lean backward a bit, taking the pressure off the dog, but remain ready to move again if he initiates another break. Some sensitive dogs will stop and back up in response to a slight lean toward them. Other dogs require much more movement from you to concede the space. For any dog, the sooner you react, the better.

Body blocks can keep uninvited dogs from jumping into your lap, as some overly friendly dogs like to do. You must use your body rather than your hands (which may cause dogs to think you are playing) to take the space away from an encroaching dog, just like dogs do to each other. Tuck your hands across your torso, and lean forward to block the dog's access to your space with the shoulder, hip, or elbow to prevent the assault. It's best to straighten up immediately once the dog has moved backward. Most dogs won't give up the first time and may come around to your other side, so you'll likely need to repeat these body blocks several times to be effective.

Body blocks can take a variety of forms: leaning while standing or seated, shuffling toward the dog, goalie-like guarding behavior which involves side-to-side movements, and stepping across the dog's path. The commonality of all these movements is that they allow people to gently control their dogs' movement and behavior by managing the space around them. Dogs respond well to body blocks, presumably because this space management is so natural to them.

**It has become popular to object to body blocks by people who consider them unkind, but I disagree. For socially healthy, behaviorally well-adjusted dogs, they are a great way to communicate. I believe that communicating with our dogs using biologically relevant signals makes it easier to make ourselves understood, which is both a kindness to them and is good for our relationship. It is true that very fearful dogs don't take well to body blocks, but they also often don't take well to petting or to seeing people stand up. That doesn't mean petting and standing up are inherently unkind.**

## Dogs as Holiday Visitors

***December 17, 2013***

Feeling panicky, I caught my dog's Kong in mid-air as it flew toward a set of porcelain miniatures and breathed a sigh of relief. Hmm. Perhaps I had been a bit smug in thinking that bringing my dog Bugsy along for the holidays would be stress-free.

He was social, had an excellent Stay, came when called, had no history of food thievery, and would not lift his leg indoors, even on a tree, so my confidence in his visiting skills was high. However, I failed to foresee that tossing his Kong in the air to get the treats out at my in-laws' house, with its collectible figurines, crystal, and teacups, would be problematic.

Whether people fully anticipate the trouble ahead, or, like me, they are surprised, bringing a dog to another's home for the holidays can cause stress. However, there are ways to improve your chances of being gracious guests.

Consider whether bringing your dog will work for everyone. Request permission first because not everyone wants a canine visitor. Even dog lovers appreciate the advance warning that allows them to, for example, put away the Ming vase displayed at the precise height of your Great Dane's perpetually wagging tail.

If your dog isn't welcome or is prone to serious transgressions such as biting, destructive chewing, or house soiling, make alternate plans.

Train your dog. The basics of good canine citizenship are Sit, Stay, Come, Leave It, and polite greetings. I also recommend teaching your dog at least one showy trick that allows him to exude charm. One client's dog jumped up on her father-in-law, but his social gaffe was immediately forgiven when she gave the cue "You goofed," and he responded by lying down and covering his face with his paws, as though in embarrassment.

Help your dog be at his best. Exercise him and provide chews, toys, and puzzles to minimize destructive chewing and counter surfing, which tend to worsen when dogs are bored or full of pent-up energy. Put temptation out of your dog's reach by volunteering to take out the trash. Don't set up your highly food-motivated dog to fail by leaving him alone, even for a minute, while the turkey is on the table.

Understand that training doesn't automatically transfer to new contexts. Just because your dog has a solid Stay at home doesn't mean he can do one at Grandma's house with a dozen cousins present. Even a couple of 5-minute training sessions to practice your dog's skills after you arrive can significantly improve his performance and manners.

Training skills aren't the only ones that may get worse when traveling. Many dogs who are completely trustworthy when left alone at home become stressed or scared when left alone in a new place, which can result in house soiling, destructive chewing, or excessive barking. Adjust your plans and expectations accordingly.

Following these suggestions may mean that this holiday is remembered as a joyous occasion instead of as the last one in which you were allowed to bring your dog.

**Bringing dogs along on holiday visits with family can lead to extra joy but also extra stress. Bugsy was always welcome on family visits, but it took a lot of effort to make it so.**

# Practical Tricks to Teach Dogs

## *March 18, 2014*

We often want our dogs to move their bodies in particular ways at the vet, at the groomer, or in social situations. Unfortunately, even if we have their best interests in mind, our dogs have no way of knowing that. Rather than grab them, lift them, or push them around—however gently—it causes less stress if we are able to tell them what we want them to do and have them perform the behavior on their own.

That's where tricks come in, because many offer more than just captivating performance opportunities.

**Crawl.** I used to ask my dog to do this when I lived on a farm in Wisconsin and he got muddy walking through the fields. Depending on the season, I would have him crawl through snow or over grass to clean up a bit before we went inside.

**Shake.** I ask dogs to present a paw on cue when cutting their nails, or at the vet to make a blood draw easier. It's also useful for checking for pad injuries or toweling off wet paws after a walk.

**Belly up.** This cue tells dogs to lie on their backs with their legs in the air. Most often, I ask a dog to do this to get a belly rub, but it can also help a veterinarian perform an examination more easily.

**Beg.** This is another way to get a dog to expose the belly, and I like to ask a dog to beg when I need to pull off burrs or seeds. If a dog's balance is good, Beg can also be a way to position a dog for a quick brush of the belly fur.

**What?** This is a cue to cock the head, which makes photos of dogs especially endearing. The less time you spend trying to get dogs into specific poses, the more likely they are to have nice expressions on their faces.

**Bow.** Having a dog perform a play bow on cue is more than just a cute trick; it can also help some dogs who are a bit awkward or nervous around other dogs. The play bow is a social signal that means, "What follows is playful in nature," and performing one near other dogs can mean that social interactions start off on the right paw, easing tensions caused by confusing or unexpected behavior.

**Up.** I use this cue to tell dogs to jump or step onto something. It's a great way to get big dogs to stand on the vet's scale. (It's even more useful when combined with a solid Stay.)

**Dry off.** This cue, which tells dogs to shake their whole body, is a way to have dogs shake off excess water after a bath or before coming inside following a walk in the rain.

Tricks involve dogs putting their bodies (or at least parts of them) in all sorts of positions, and that variety of movement and behavior is what gives tricks such practical applications. The entertainment value is just a bonus.

**I absolutely love trick training! It's the best possible combination of recreational and practical. How often in life is something that's this fun also so useful?**

## Love Has No Age Limit

***July 22, 2014***

The following advice is from *Love Has No Age Limit: Welcoming an Adopted Dog into Your Home,* a book that Patricia B. McConnell, Ph.D. and I co-wrote to help people with the transition from "new dog" to "family member." It offers guidance through the first steps of new relationships with adult or adolescent dogs. Though it's difficult to be specific because every dog is different, here are some general guidelines.

You may be prepared to enter into a close and emotional relationship with your new dog the day you bring her home, and you probably have expectations of who she should be, but your adopted dog has no such mindset. After all, you probably spent a long time thinking about adopting, while your new dog had no idea what to expect when you picked her up. Everything, including you, is new to her.

No matter what their background, adult dogs are similar to adult people in that they often need time to form social relationships. Young pups, with rare exceptions, will instantly bond to their new humans. But adult dogs vary tremendously; some will lick your hand, kiss your face, and not want to leave your side within minutes of meeting you. Other dogs are like people involuntarily thrown into a blind date, and have no concept of how to relate to you until they get to know you better. That doesn't mean they won't become your best friend, it just means that the level of closeness you reach in the first few days or weeks might be different than if your dog was an 8-week-old pup.

As well as needing time to get to know the people in the household, adult dogs need time to acclimate to everything else that is new to them. It's important in the first few days to balance giving your new dog attention with giving her time and space to become comfortable in a new home and with everything she's encountering in it.

Spend as much time as possible with your new dog in the first few days. It's only by spending time together that you can learn more about your dog's habits and personality, and doing it early on will make adjusting to life together smoother for both of you.

Be respectful when you are getting to know a new dog. Keep your face away from the dog's face, don't hug or grab at the dog, and avoid letting young children interact without carefully watching for signs of discomfort from the dog. Don't throw the dog in with other dogs in hopes it will all work out. Be careful and don't push your dog into dealing with too much too soon.

See the dog, not the story. What your new dog needs most of all is the same thing a person needs—to be accepted for who they are, rather than to be labeled. Do all you can to see her for who she is *now*, not who she was years ago or who you think she should be.

**In the later stages of writing this book, Trisha and I had a discussion of some of the things left to do. At one point, I mentioned that I really, really wanted to use the phrase "Love has no age limit" somewhere in the book. At another point in the conversation, Trisha expressed her concern that we did not yet have a title for this booklet, which we just kept calling "the rescue booklet." It was pretty late in our conversation that we put these two statements together and resolved the two problems simultaneously.**

## Treating Kids Like Dogs

### *August 19, 2014*

I've spent many years observing canine behavior, interacting with my four-legged best friends recreationally and professionally, and influencing what dogs do by using positive training techniques. It was only natural that these career skills and experiences influenced my approach to parenting.

At first it was not a conscious effort. It's just that after years of working with dogs, I knew far more about raising that species than I did about raising children. Unsurprisingly, I sometimes treated my babies like puppies.

When I wanted to get my baby's attention, I made smooching noises, clapped, or said, "Whoop Whoop," because those sounds all work on dogs. When my oldest was about 10 months old, someone asked me if he was walking yet. I answered that he was often up on his hind legs and holding onto the couch instead of giving the more species-appropriate reply that he was at the cruising stage. Later, I used the terms "toilet training" and "house training" interchangeably, to the embarrassment of friends and family.

It wasn't long before I began actively applying my dog training and canine behavior skills to parenting. For example, I like to keep both kids and dogs right near me when there is any danger from cars. In the dog world, that's what leashes and heeling accomplish. Once my kids were old enough to toddle, the rule was that whenever we were in the street or in a parking lot, they had to hold my hand. This compares to having your dog on a leash. When they got a little older (around 3 or 4 years) old they graduated to "Stick Close." which looks a lot like heeling. In fact, the only reason I called it "Stick Close" instead of "Heel" is because I didn't want the other moms or the neighbors giving me weird looks.

Those weird looks were unavoidable when I applied canine boundary training as a strategy for keeping my kids safe from cars. To teach dogs to stay in an area without a fence involves having them practice stopping at the boundary over and over, so I did the same thing with my kids. Every day, we practiced running down the driveway and stopping at the curb. My hope was that when they were chasing after a ball some day while I was distracted for the proverbial second, their habit of stopping at that spot would prevent them from running into the street and therefore protect them.

Don't be alarmed, but I even taught my kids to play fetch with each other. In my defense, fetch is good exercise, it's a cooperative game that teaches them to take turns, and everyone deserves to have interesting childhood experiences to discuss with a therapist. (My kids will also be able to share that they were taught Stay as a specific skill.) Plus, when I tell people I've borrowed techniques from my dog training and canine behavior work to use on my kids, I assume this is what they imagine is going on at our house, anyway.

**The longer I have been a parent, the more I realize that my dog training skills apply to children and to adults. That eventually led me to write the book *Treat Everyone Like a Dog: How a Dog Trainer's World View Can Improve Your Life,* which is dedicated to my sons and to my husband.**

## Controlling Canine Destructive Chewing

***November 4, 2014***

Leaving a dog alone and unconfined at home works out beautifully for millions of families. These people go to work, and return to a dog who is pleased to see them and a house that is exactly as it was hours ago. However, others return to a dog who is pleased to see them *and* a mess courtesy of powerful canine jaws.

Destructive chewing is a common problem, especially for puppies and adolescent dogs. Retrievers and terriers are famous for chewing, but it can happen with any dog. I knew a Great Dane who actually *ate* a couch, and a Chihuahua who destroyed an entire wall. Handling destructive chewing requires a combination of tactics: dog proofing the house, confining the dog, minimizing boredom, and providing appropriate outlets for chewing.

Dog proofing the house is a form of prevention that's important for the safety of both dogs and houses. Make sure tempting items—shoes, plants, children's toys, and electric cords—are out of reach. Keep food off counters, remote controls off the sofa, and anything leather such as gloves and purses hidden in drawers. Use baby locks on cabinets to prevent access to valuable or dangerous items. Deterrent sprays with a citrus scent (which many dogs find repulsive) may inhibit chewing on table legs and rugs.

Confining dogs to puppy-proofed areas is an important strategy for managing destructive chewing. Options include the laundry room, a bathroom, the garage if temperatures are not too extreme, or a crate that's big enough that the dog can stand up and turn around. Though most of us eventually want our dogs to have the run of the house, it's wise to confine dogs temporarily until they are old enough, mature enough, and well-trained enough to handle that freedom. It is essential, though, for them to have ample time free from confinement to exercise and have fun.

Many dogs chew out of boredom, so the more exciting dogs' lives are, the less likely they are to chew. Some dogs chew no matter what, but some only do it when they've got nothing else going on. To enhance any dog's life, the formula is plenty of physical exercise such as walks, runs, swimming, and play; lots of mental exercise in the form of adventures to new places, training classes, or learning new tricks; and making them work for their food. Have them find their food scattered all over the house or the yard, or have them extract it from Kongs, Goodie Balls, Buster Cubes, or Roll-a-Treat balls, rather than simply eating from a bowl.

So many dogs have such a strong natural drive to chew, but chewing itself is not the problem. Chewing on inappropriate things is. Dogs need to have things they can chew on such as bones, Kongs, rawhides, Nylabones, and Greenies. Check with a veterinarian to determine which chew items are appropriate for your dog, and which of them require supervision.

These ideas can help you create a happy, peaceful coexistence between your dog and your house.

**Rereading this many years after writing it, I realize that I included rawhides and bones as options without pointing out that due to the risk of choking, dogs should always be supervised when enjoying these items.**

# How to Have Animal Magnetism with Dogs

### *July 21, 2015*

We all know people who have a magical way with dogs. Dogs flock to them, choose them out of a crowd, and lovingly settle at their feet. How do these "Pied Pipers" do it? What do they know that most people don't?

Due to specializing in aggression and other serious canine behavioral issues, I regularly work with dogs who, to put it mildly, don't make friends easily. I have to gain their trust as well as help them relax and warm up to me enough that I can work with them while keeping myself (and my clients) safe. I have to be a Canine Pied Piper to do my job effectively. I want to be popular with dogs outside of work, too, so I do the same things in social settings that I do professionally. These techniques are effective whether a dog has serious behavioral problems or is just like Lassie.

Anyone can adopt these behaviors to become the type of person dogs are drawn to, but they don't come naturally to most people. Being a Canine Pied Piper involves behaving less like typical people would and more like dogs do. There are three basic components to achieving canine animal magnetism, and the true Pied Piper does them all.

First, Pied Pipers present themselves properly to make a good impression and avoid appearing intimidating. Many dogs get spooked by anything that changes a person's silhouette. That's why going without hats, backpacks, or sunglasses and not holding anything (books, a clipboard, grocery bags, a toolbox, a child) make people more likeable and appealing to dogs.

Second, Pied Pipers greet dogs appropriately. When people greet each other, they tend to face forward, approach directly, lean towards the other person, make eye contact while extending a hand to shake, and bellow out a hearty hello. Though polite in human society, this behavior is rude and threatening in canine society. To endear themselves to dogs, Pied Pipers greet dogs the way that dogs greet each other. They turn slightly to the side, lean slightly away from the dog, and get low by squatting if it is safe to do so. They don't stare, they don't reach out towards or touch dogs, they speak softly, and they let dogs approach them.

Third, Pied Pipers use classical conditioning to teach dogs to associate them with feeling good so that dogs want to be around them. Before dogs have a chance to experience any negative emotions in their presence, Pied Pipers make sure the dogs feel good. Whatever they love most works best. For most dogs, a little slice of heaven comes in the form of treats such as steak, chicken, or freeze-dried liver (no dry biscuits have the power!), but some dogs prefer balls or squeaky toys, and for the occasional dog, petting will be most effective.

By presenting oneself properly, greeting dogs politely, and conditioning dogs to feel good, anyone can be a Canine Pied Piper who appears to have a magic way with dogs and is beloved by them.

**I am using the term "Pied Piper" here in the positive sense of being charming rather than the negative sense of causing harm or manipulating others for their own gain. When I wrote this, I was unaware of both the negative connotation and the historical facts about the Pied Piper of Hamelin. Mea culpa.**

## Benefits of Adopting an Adult Dog

***August 18, 2015***

The charm of puppies is undeniable and though there are many reasons to adopt one, there are plenty of compelling reasons to welcome an adult dog into your life instead.

**Freedom from size-surprise.** Years ago, a woman in one of my training classes asked me what breeds might be in her puppy. The shelter had told her that he would be less than 50 pounds, but she was suspicious. When I told her the only breed that seemed obvious was Great Dane, she went quite pale and had to sit down. Her dog grew to 28 inches and roughly 100 pounds. With an adult dog, you know what you're getting.

**House training may be a fait accompli.** Teaching a dog that the house in general, and the fancy Persian rug in particular, are not bathrooms involves vigilance, schedule adjustments, sleep deprivation, frustration, and at least the occasional cleanup. If your dog comes pre-trained, be grateful.

**Lower exercise requirements.** All dogs need exercise, but puppies typically need 45 minutes of off-leash exercise twice a day plus lots of small walks or activity sessions. In contrast, most adult dogs thrive with a single daily session of good exercise.

**So many dogs need families.** Canine overpopulation remains a serious problem. When you choose an adult, you are providing a loving home to an individual who might not otherwise be adopted.

**A possible pass on typical puppy issues.** It's not easy raising a puppy, and anyone who suggests otherwise is not to be trusted. Puppy biting and destructive chewing are perfectly normal behaviors, but nobody thinks they are positive aspects of dog ownership. How lovely to adopt a dog who has already passed through some of the most trying developmental stages!

**Already spayed or neutered.** I strongly endorse these procedures, but that doesn't mean they are fun for anyone. There are occasional reactions to the surgeries, the difficulty of keeping a puppy calm for a few days afterwards can be a major headache, and it's more fun to spend money on toys, treats, and dog beds. If your dog is not yet spayed or neutered, by all means get it done, but if someone already took care of it, consider that a bonus.

**Training experience.** Many adult dogs already know how to Come, Lie Down, Stay, Shake, or perform various other behaviors on cue. Of course, there are also adult dogs who have apparently never even heard the word "Sit" before, much less learned to do so when asked. Still, there's a good chance an adult dog has had at least a little training.

**Feeling the love.** Dogs whose pasts include less-than-perfect situations are likely to appreciate being in a loving home. The common wisdom that strong bonds are formed only by acquiring a dog in puppyhood is just not true. Many re-homed dogs are highly affiliative, super-attached companions, and the love of someone's life.

No matter what the age of the dog you welcome into your heart and into your home, I wish you much happiness together.

**Patricia B. McConnell, PhD and I co-wrote *Love Has No Age Limit: Welcoming an Adopted Dog Into Your Home* to make the whole experience of adopting adult dogs just a little bit easier.**

## Follow Etiquette When Encountering a Service Dog

***July 19, 2016***

Service dogs enhance lives, but those who have them often struggle with the behavior of other people. Following some basic guidelines can relieve this burden. Here's what many human members of service dog teams wish everyone knew.

1. Service dogs are working dogs, not pets. They have a job to do, and they can't do it effectively if they are distracted. Just as you would not massage a person who

is laying tile or initiate a game of catch with someone who is directing traffic, don't pet or play with service dogs. It's best to politely ignore them.

2. People with service dogs need a dog because of a disability, which is why it's unkind and inappropriate to voice your desire for a service dog. If you say, "I wish I had a service dog," it sounds like you're wishing for a disability, too.

3. There are no specific certifications for service dogs. To qualify, a dog must be trained to perform a specific task or tasks that are directly related to the person's disability.

4. Service dogs don't have to wear identification such as vests or tags. Many people choose to have their dogs identified as service dogs in that way, but it is not required.

5. The right of service dogs to have access to any place that is open to the public is protected by law. Businesses may ask only, "Is that a service dog?" and "What task or work has your dog been trained to perform for you?" They may not ask about your disability or the exact function the dog serves. It's beyond rude to say, "What's wrong with you?" or "How come you need a service dog?" or "You don't look disabled."

6. Service dogs typically have happy lives, so don't worry that they work too hard and have no fun. They have opportunities to be off duty, which is when they can sniff around and play. Sure, they work, but they have full lives, and share a close relationship with their people.

7. Service dogs perform many tasks beyond the best-known work of guide dogs for the blind. Service dogs may alert people with diabetes to dangerously low blood sugar levels or a person with epilepsy to an impending seizure. They may let people who can't hear know that the doorbell has rung or an alarm has gone off. People with PTSD have service dogs to alert them to someone approaching from behind, to block people from approaching too quickly or too close, and to check a room for people before the handler enters. Service dogs may offer physical support to people with balance issues, and they may retrieve items or turn lights on and off for people whose disability prevents them from doing so.

Be respectful by letting these dogs do their jobs and by refraining from asking nosy questions, no matter how curious you are. Becoming well-versed in service dog etiquette is a kindness to the people who rely on these dogs.

**I am happy to report that in the years since I wrote this, people are far more aware of the proper way to treat both members of service dog teams.**

# Giving Thanks for Canine Companions

*November 22, 2016*

Like many people, I'm thankful today and every day for the animals—especially the dogs—in my life. Because of our canine companions, we are happier and healthier in so many ways. Here are some excellent reasons to be thankful for them.

A recent study showed that children who live with animals are more likely to develop strong immune systems than kids without such exposure. They are also less likely to develop asthma or to become allergic to animals.

Dogs affect our physiology in a number of positive ways. Petting dogs (or playing with them) raises our levels of oxytocin, which is the same hormone that induces feelings of romantic love and plays such a large role in the bonding of mothers and babies. Interacting with dogs also lowers the level of cortisol, which is strongly associated with feelings of stress. Perhaps the reduction of stress partially accounts for the lower blood pressure and decreased levels of cholesterol and triglycerides of dog owners compared with their dog-free peers.

These physiological benefits of dog ownership may explain why people who own dogs have better recovery rates following heart attacks than people without dogs, but exercise may also be a key factor in the enhancements to our health due to dogs. Dogs get us out of the house. They need exercise, and that means that in weather that is foul or fair, many dog owners head outside, like it or not. Dog walking decreases the likelihood of obesity and keeps us fitter than we might otherwise be. Our dogs' need for exercise also keeps us in touch with our friends and neighbors. It's common to walk dogs with friends, or make new ones who are also out walking their dogs.

Pets make great social facilitators, according to a variety of scientific studies. People pay more attention to people with dogs and are more socially engaged with them. Bystanders are more helpful to people in need of assistance who have a dog with them. Dogs may even make you more attractive. In one study, a man asking women out had more success when he was with a dog than when he was alone. For the sake of science, one man asked 240 women for their phone number. Of the 120 times he did this without a dog, 11 out of 120 women (9.2 percent) said yes, but 34 out of the 120 women (28.3 percent) who heard his pitch when he had a dog were charmed enough to comply with his request. This threefold increase in success shows that "wingdogs" are superior to wingmen!

There are so many people who feel better around dogs, whether that means they feel less lonely, they have a sense of purpose, or they have a reason to smile. It's common for those in the worst throes of grief to give their dogs credit for the few laughs that they have while in such pain. Still others are made to feel whole by their dogs, feeling as though they are literally not complete without them.

**There was a time when many people felt too embarrassed to express the importance of their dogs in their lives, but thankfully that's not so common now. With scientific research showing that dogs add value to our lives and keep us healthier, there's no shame in acknowledging how much we need them and rely on them.**

## Why the Circle Wag Is the Friendliest One of All

***March 20, 2018***

The simple view that a tail wag always means a dog is friendly and happy is an inaccurate interpretation of the tail wag. The complexity of this behavior is immense, and misunderstanding a dog's tail wag can lead to confusion or even danger. Sure, friendly dogs tend to wag their tails, but so do dogs who are not friendly, including many who are aggressive.

However, there is one type of tail wag that almost always means what most people think a tail wag means—that the dog is happy and friendly. That type of wag is called by the descriptive term "circle wag," meaning that the tail is moving at the base and that the tip is going around in a broad circle.

Most often, dogs will circle wag if they are particularly overjoyed to see someone special, especially someone they haven't seen recently. Many videos of reunions between military people returning from deployment and the dogs who stayed behind feature circle wags. Some dogs have this level of enthusiasm—and the corresponding circle wag—every time they greet members of their family no matter the length of the absence.

Though it is generally correct to interpret a circle wag as a sign of friendliness, even that is an oversimplification. My late dog Bugsy illustrated this. He loved to play with most other dogs and generally had good social skills when off leash. However, his on-leash behavior was problematic. Until I worked extensively with him to modify his behavior, he would bark, growl, and lunge whenever he saw a dog. Additionally, I had to work very hard to manage him around toys or food because his natural inclination was to be possessive of anything he considered valuable, and to act aggressively towards a dog or a person around his treasures. To add to his list of issues, he exhibited signs of fearfulness in many situations.

Bugsy would circle wag when charging full speed toward other dogs who came onto the property when I lived on a 150-acre farm in Wisconsin. Before running at them, he would bark and stiffen—not signs of friendliness by any stretch of the imagination—but when he reached the dog, he would initiate play after a tense greeting. It is one of the few times I have seen a dog exhibit a circle wag when not in an obviously friendly or happy state, but he really did like playing with dogs, so maybe he was anticipating that. My guess is that he was in an ambivalent state—open to the possibility of a friendly encounter, but nervous about the other dog at the same time.

Some dogs wag their tails in circles regularly in response to any positive experience. Other dogs only do it at specific times, such as when greeting their best dog buddy, when their owner comes home after being away for a long time, or when they realize it's time for a walk. Like most behavior, the more we know about it, the more complex we realize it is.

**The circle wag is also sometimes called the helicopter wag or the propeller wag.**

## Why Premature Graying in Dogs Matters

***May 22, 2018***

Humans go gray at a young age due to a variety of influences, including genetics, disease, and stress. Many parents swear their kids are to blame, but puppies are probably not the reason for premature grayness in canines. The 2016 study "Anxiety and impulsivity: Factors associated with premature graying in dogs" by Camille King, Thomas J. Smith, Temple Grandin, and Peter Borchelt in the journal *Applied Animal Behaviour Science* suggests that a dog's tendency to go gray may be correlated with multiple issues, many of them emotional in nature.

For the study, four hundred 1- to 4-year-old dogs were photographed from the front and side so that the degree of graying on their muzzle could be assessed. They were scored 0 = no gray, 1 = frontal gray, 2 = half gray and 3 = full gray. Dogs were only included in the study if it was possible to determine how gray their muzzles were. (The 43 dogs who were white or had merle coloring were excluded.)

The owners of the dogs filled out a 42-question survey. Data on anxious behaviors, impulsive behaviors, fears, size, age, sex, number of dogs and cats in the household, time spent unsupervised outdoors, whether they were spayed or neutered, medical issues, and participation in organized sports or activities were collected. The people who evaluated the photographs had no knowledge of the questionnaires, which prevented accidental bias in their judgements of the degree of graying.

The survey was designed so the owners were unaware of the study's purpose. (They were told it was a study involving dog lifestyle.) In addition to questions probing the factors of interest, there were distractor questions to prevent people from biasing their answers based on what they thought researchers were investigating. Distractor questions included, "Does your dog have hind limb dew claws?"

Researchers found an association between graying on the muzzle with anxious behaviors and with impulse behaviors, as well as with fears of loud noises, unfamiliar people, and unfamiliar dogs. The extent of grayness was positively correlated with age, and female dogs were more gray than male dogs. They found no link between premature grayness and size, being spayed or neutered, medical problems, reactions to thunderstorms, fear of unfamiliar places, number of dogs or cats in the household, time spent outside unsupervised, or being involved in organized activities.

The increased understanding of premature graying provided by this research opens up possibilities for helping dogs. Being anxious or fearful or struggling with impulse control is hard on dogs, and it is beneficial for them to receive help to overcome these challenges. Premature graying can tip people off to the possibility that a problem may be present, making intervention more likely to happen and to happen sooner. If behaviorists, veterinarians, trainers, and other dog professionals know that a gray muzzle in a young dog may indicate that he or she is dealing with certain issues, perhaps they will perform a more thorough assessment, or refer the dog to other specialists for further evaluation. The information revealed in this study can be used to make life better and easier for dogs.

**My husband's identical twin brother started going gray nearly 4 years earlier than my husband. It doesn't seem like a coincidence that his twin brother had his first child 3 years and 9 months before our oldest was born, which is further evidence that going gray is not**

**just genetic and that environmental factors are at play. American presidents famously go gray during their time in office, leading to striking before and after pictures.**

## Running Can Be a Joyful Escape for Fidos

*May 21, 2019*

When your dog can't handle the sight of another dog, every walk is an adventure, but not the fun kind. Aggressiveness to other dogs when on a leash is one of the most common problems I see in my work as a canine behaviorist and dog trainer, and I use multiple techniques to help dogs with this issue.

One effective method involves training the dog to perform a specific behavior every time she sees another dog so she does that instead of barking, growling, and lunging. Two of the most common behaviors are to look at you, and to turn around and head the other way. Dogs must begin where they can succeed such as with a calm dog far away, and work gradually towards greater challenges like closer dogs, more rambunctious ones, or multiple dogs.

Naturally, it's best to prevent situations in which a dog will fail, but I like to offer a plan for unexpected encounters. That way, my clients have something more productive to do when another dog surprises them than struggle with their own out-of-control dog and say to themselves, "This is exactly what Karen said to avoid because it could cause a setback in our training program. This should definitely *not* be happening."

I always recommend getting out of a bad situation ASAP. Running is a useful strategy because it's the fastest way to increase the distance between your dog and trouble. So, if the situation is so intense that there is no way that a dog will be able to respond appropriately, I advise people to run away. If retreat is impossible, the next best option is to run past the other dog.

One client and her dog had progressed steadily for months when we were caught off guard twice during the same training session by dogs suddenly appearing out of nowhere, causing us twice to run to escape. I felt disheartened by this bad luck until I noticed that the dog was happier than usual. When we saw the next dog at a manageable distance, we were prepared to ask him to look at his owner in order to earn treats, but he just took off running. He was running gleefully rather than to escape.

After that day, his response to dogs improved markedly. He would see another dog and look up at his owner expectantly, ready and eager to run. The opportunity to run was far more powerful in affecting his emotions than the opportunity to respond to a cue and receive treats. It was fun, so it positively affected how he felt, and therefore his behavior, around other dogs.

He had learned to be happy about seeing dogs because it meant he was going to get to run, which most dogs adore doing and consider to be the best kind of play. In the years since my experiences with that dog, I have used this technique frequently with other dogs as more than just an emergency escape, but actually as a way to improve dogs' behavior long term.

**Running makes so many dogs happy in so many situations, whether it is just a burst of speed for fun or a way to escape a troublesome situation. The use of running to help dogs is an underutilized and powerful way to positively impact their emotions and behavior.**

## Keeping Dogs Safe from Coyote Encounters

***August 6, 2019***

Coyote sightings in Flagstaff seem more numerous this year than usual. That is especially concerning for people with dogs because the potential for injurious interactions is higher when a dog is present. Coyotes are generally shy of people or disinterested in them, but they often attempt to engage with dogs, even if people are around. Large dogs may be viewed as competitors while small dogs may be regarded as prey. Some coyotes have rabies or other diseases that can be transmitted through bites. So, how do you keep your dog safe from coyotes? Though there are no guarantees about the behavior of any wild animal, there are steps you can take to make trouble less likely.

Minimize the chances of encountering a coyote at all. Keep your dog on a 6-foot leash rather than on an extendable one, and by all means reconsider letting your dog off leash. Stay on trails, taking care to avoid places with lots of brush because that may result in your dog and a coyote surprising each other. Few canines respond well to such surprises. Steer clear of locations that you know have had a lot of coyote activity, especially when coyote pups are present. Try not to be out with your dog around sunrise or sunset when coyote encounters are most likely.

Though avoidance is wise, it's important to know what to do if you and your dog are walking along, minding your own business, when a coyote comes on the scene. Obviously, don't panic, which is (just as obviously) hard to accomplish, but do your best. Most dogs pick up on our emotions and even faking being calm is better than nothing. Then, control your dog so that he or she doesn't run, as running can incite coyotes to chase. Put a leash on any dogs who are unleashed. Pick up small dogs. Hold all dogs close and do what you can to soothe them and keep them calm.

In addition to physically managing your dog, you should act to keep the coyote from approaching. The most important advice is DO NOT RUN. Running is likely to lead to a chase, and being chased by a coyote may lead to your dog being attacked by one. It is better to stand tall, maintain eye contact and not to turn your back. Try to frighten the coyote so that it will retreat. Try yelling, stomping your feet, and tossing rocks or branches near (but not at) the coyote. You can also make noise with a jacket, or shine a bright light at it. It seems cruel, but it's actually kind to maintain coyotes' fear of people. The boldest of them put their own safety at risk if they are viewed as a nuisance.

With the Museum Fire forcing many animals out of their homes, there is the possibility that coyote encounters will increase. Stay safe by avoiding coyotes when possible and by knowing what to do if you and your dog do cross paths with one.

**This column was inspired by a coyote who made us very nervous when spotted during a run with my friend Mackenzie and her dog Fiona during the summer of 2019 when the Museum Fire was affecting both the people and the wildlife in Flagstaff.**

## Dog Paws, Brains, and Tails: Sides Matter

***April 21, 2020***

Home with your dogs a lot lately thanks to the pandemic? Interested in conducting experiments on them? (Obviously, I mean nice ones that don't hurt!) An easy observational study is determining whether your dog is right-pawed, left-pawed, or ambilateral (no paw preference). Dogs are evenly split between these three groups. Before you figure out where your dog fits, here's why it matters.

Paw preference reveals whether one side of the brain is more active than the other. Each side of the body is controlled by the opposite side of the brain, so a preference for the right paw reveals a more active left side of the brain while a left-paw preference indicates more activity in the right side of the brain.

A disparity in the activity levels matters because each hemisphere of the brain plays a different role in emotional processing. The function of each side is still being explored, but overall, the left side does more processing of positive emotions such as happiness, excitement, and affection as well as the response to anything familiar. The right side deals more with negative emotions like fear and sadness as well as the response to novel items. Paw preference is linked with emotional tendencies and therefore with behavioral problems because emotional inclinations play a big role in the development of problems.

Left-pawed dogs are more likely to have stranger-directed aggression than right-pawed or ambilateral dogs. Being right-pawed is associated with being calmer and showing lower arousal levels in response to strangers and to new things. Ambilateral dogs have been reported to be more reactive to loud noises than dogs with a paw preference. In one study, dogs with stronger paw preferences were more confident, less cautious, quicker to relax or become playful in new environments, less prone to arousal and anxiety, and acted more calmly in response to novel stimuli and strangers than dogs with weaker paw preferences. There is a higher success rate for right-pawed dogs than for left-pawed dogs as guide dog candidates, who need to be emotionally stable and behaviorally calm to succeed and become guide dogs.

Tail wagging can show a side bias based on the brain hemisphere acting most strongly on its motion. When a dog's tail wags higher to the right, that's an indication of a positive emotional state because the muscles on the right side of the body (controlled by the left side of the brain) are activated. If the tail is wagging higher to the left, the dog may be showing signs of a negative emotional state.

To assess your own dog's paw preference, observe your dog extracting food from a Kong. Every time your dog uses a paw to stabilize the Kong, record which paw was used. If the dog uses both paws simultaneously, record that. After 100 instances of using one or both paws to stabilize the Kong, you can determine your dog's paw preference (if any) as well as the strength of that preference.

**Ever since a 2007 paper came out about tail-wagging asymmetries, dog trainers and canine behaviorists have been awkwardly trying to line themselves up properly relative to the dog's body to see if it's possible to detect differences in the amplitude of the tail wag to each side.**

# How to Introduce Dogs and Cats the Right Way

*June 23, 2020*

Adopting new dogs is as popular in 2020 as searching for toilet paper and baking sourdough bread. Newly adopted pets must meet all the members of the new household, and the trickiest introductions frequently involve cats. Regrettably, the most common way to introduce a dog and a cat consists of putting them in a room together and hoping for the best. That all too often results in the dog chasing the terrified cat who then never wants to come out of hiding. This can even happen with a dog who loves cats, wants to play, and would not physically hurt the cat. There is a better way.

**Keep them separated at first.** For the first few days or weeks, keep the cat in her own area of the house with all she needs to be happy—food, water, litterbox, toys, and good sleeping spots. Spend lots of time with her so that she does not become isolated. Allow the dog and the cat to get used to the sight and smell of one another with a gate between them. Trade their blankets so they get used to each other's scent. Encourage positive feelings by giving them both lots of toys, treats, and attention. Teach the dog to sit or lie down when seeing the cat.

**Control interactions.** Once they are calm and comfortable in the presence of one another, and the dog willingly sits or lies down at the sight of the cat, begin controlled interactions without a gate, but with leashes.

Ideally, have one person for each pet. Begin with short sessions (less than 5 minutes) and always keep their experiences positive with treats, petting, toys, and a calm manner. Do not try to get them to be close together—just keep them happy in each other's presence. Do several sessions a day and gradually make them longer.

**Provide high perches for the cat.** Provide a high spot in every room the cat can access but the dog can't. Cat condos, cat trees, and shelves are options, but so are high bookcases and the fridge. These function as escape routes for cats where they can feel safe and get away from a dog for any reason. To encourage the cat to use these high spots, place cat treats on them.

**Use artificial scent marking.** Use Feliway®, which is an artificial feline scent marking product that helps cats feel comfortable, even in new environments that might otherwise upset them. Put a little in the high spots for the cats as well as any common perches or sleeping spots.

**Do it right because you can't do it over.** If an initial dog and cat introduction doesn't go well, it will be really hard to turn that relationship around. Dog-cat introductions are very different than dog-dog ones because rebounding from a rocky start is much harder with the former. Introductions between dogs and cats should be done the right way to maximize the chances they will be friends, or at least tolerate one another.

**Though I wrote this article months before the 2020 election, it was read far more after then President-**

**elect Joe Biden and Dr. Jill Biden revealed they planned to add a cat to their household. The cat would be joining the Biden's two dogs, and many people became interested in how to make dog and cat introductions successfully.**

## Training Animals to Participate in Their Own Health Care

***April 20, 2021***

If you need to draw blood for medical purposes from a hippopotamus (an animal with skin thick enough to break needles), you can either train that individual to stand while you lift up the tail and take blood from the soft skin underneath, or you can knock it out with anesthesia. There are clear benefits in terms of safety, efficiency, and cost of the former, which is why zoos teach animals various skills and behaviors that make their care easier and better.

Professionals in the zoo community know that training isn't a luxury, but an essential part of animal care. In zoos, it has been common for decades to train animals to participate in their own care. That makes it easier, safer, and less expensive to meet their medical and grooming needs without force, stress, or anesthesia. Similarly, many dog trainers teach dogs to perform such tasks as presenting their paw for a blood draw, and for the same reasons.

It's unrealistic to think you can force giraffes to step on the scale, so they are trained and reinforced for participating in this useful behavior. This is often done by teaching the giraffe to "Target" which means to touch a part of its body (often the nose) to an object like a foam ball on the end of a pole. By positioning the pole where you want the animal to go, you can guide the animal to move to a certain spot. Targets are also used to move animals to different parts of their exhibit, or even to another area of the zoo. Though it's physically possible to force a dog onto the scale at the veterinarian, it's better to train dogs to do it themselves willingly with the use of a target.

Captive elephants require regular nail trimming to prevent overgrown nails from cracking, splitting, and becoming infected, which can lead to pain and abnormal gaits. Animal trainers teach them to present their feet to receive this necessary grooming. Many dog trainers also routinely teach dogs to offer their paws for nail trims to avoid the antagonistic interactions that might otherwise occur.

Zoo animals as well as dogs can be trained to open their mouths or present various body parts for inspection, to lie on their backs to allow medical staff to probe their bellies, and to take their medicine. They can be taught to stand still to accept eye drops, ear drops, or the application of any sort of ointment, and even to give a urine or fecal sample upon request.

Anything we can teach animals to do that eliminates the need to force them to do something, or to surprise or upset them, makes their lives better. Teaching an animal to engage in an interaction or behavior voluntarily is better for their mental and emotional health as well as safer for them and for the people involved. Dog trainers have learned a lot from those who work with zoo animals, including how to care for them while minimizing stress and conflict.

**Husbandry training (teaching animals to participate in their own care) is gaining popularity in the dog training world, which has made grooming, medical exams, and medical treatment easier, far less stressful, and less dangerous for dogs and people alike. My favorite program for this husbandry training is Laura Monaco Torelli's, which she launched in 2007 and is called *Ready, Set, for Groomer and Vet.***

## Using Reinforcement to Train Dogs to Come

***November 16, 2021***

When I call a dog to come, I mean, "It doesn't matter if you see a herd of 20 deer and each one is offering an antler or a thigh bone for you to enjoy. You are still supposed to turn your furry little self around and run to me." For a dog to behave as though the cue means essentially the same thing to them, I need the dog to hear "Come!" and think, in some canine kind of way,

"Excellent! Whatever's going on over here, she's got something better over there."

Dogs who know from lots of previous experiences that hearing that cue predicts great things are going to want to head over to you—and quickly. A great deal of practice with high-quality reinforcement in which a dog associates responding to "Come!" with positive events is necessary to train a reliable recall.

Consider the behavior of dogs when they hear a treat bag rattling or see someone pick up their leash. These dogs come running because they know that something wonderful will happen when they do—they'll receive treats or go for a walk. The basics of teaching a good recall involves teaching dogs to be as happy about hearing the word, "Come!" as they are about those other predictors of good things, and that only happens if your cue does predict good things.

A dog can only receive reinforcement for coming in the form of treats, a new toy, a bone, a chance to chase you etc. if they do come when called, so it's important in training to organize situations that make it very likely dogs will respond correctly. Set dogs up for success at every stage of the training in settings where it will be easy for them to choose to run to you. At the beginning, that means doing so when not much else is going on, so they are not distracted.

A first step might be standing five feet away from a dog in the living room. A dog who comes a short distance and receives chicken is learning that coming when called is fun, and so very, very worth it. That is 100 times easier than calling them to come at the park when dogs, squirrels, birds, and plenty of delicious smells make it way too distracting for a dog to come unless they have been specifically trained to do so in that context. You must work up to those more difficult situations with a series of increasing challenges, lots of repetition at each stage, and outstanding reinforcement throughout the training process.

If you call a dog away from a fun play session with a best buddy but reinforce with something so-so like a dry biscuit, they could learn that it's not worth coming to you, so always reinforce well. Training your dog to come by consistently reinforcing them for doing so with items of high value is a way to insure they're willing to come and be oh-so-glad they did—every single time.

**It's wise to be even more generous when training recalls than when training any other skill. I don't reinforce every time with other cues except in the early stages of teaching a dog something new. Training dogs to come when called illustrates much about my approach to dog training. I want it to work for the dogs and for the people. Using high quality and generous reinforcement, we can align what the people want the dogs to do with what the dogs want to do, so it's a win for everyone.**

# Acknowledgments

Creating a collection of columns and turning them into a book is long-term project that involves many individuals. I'm grateful to a lot of people who taught me so much about dogs and how to work with them.

Patricia McConnell, PhD took me "under her paw" as an assistant dog trainer in 1997 while I was still in graduate school, and gave me a full-time job as her aggression specialist in 2000 after 6 months of interning with her the year before.

Training director Aimee Jarosz and trainers Chelse Wagner, Julie Vanderloop, Shannon Skolaski, Kim Keen, and all the other Dog's Best Friend trainers were generous with their knowledge, and I learned from all of them. Their lessons still inform what I do and how I do it. Pia Silvani and Trish King, whom I met at conferences, had innovative ideas that I incorporated into my own work. These women inspired me to keep seeking better ways and new ideas all the time.

My family has supported my love for dogs as long as I can remember. My parents, Bobbi and Ralph London, let me visit the Old English Sheepdog across the street when I was three years old, hang out with the mixed breed next door, and participate in the training of my aunt and uncle's Great Danes. Years later, the support continues. Dad edits each of my columns before submission (as did Mom while she was alive), offering suggestions about content, grammar, and organization. My sister Marla London has patiently listened for decades to my eternal talk about dogs. Furthermore, she lent her excellent writing and editing skills when reading a draft of this book.

My husband, Richard Hofstetter, and our sons, Brian Hofstetter and Evan Hofstetter, willingly welcomed so many dogs into our home whether I was training them, fostering them, or dog sitting them. They all helped me with many clients over the years whether I needed a man, a child doing a handstand or one crawling on the ground, a runner, a skateboarder, or a biker to help the dogs I was working with to overcome their fears, or needed visitors at the door to help teach dogs to sit instead of jumping up. They were all also forgiving whenever my "I have a deadline!" personality made an appearance.

My oxytocin levels rise when I see the photos on the cover of this book, and my heart glows with gratitude for the people who allowed me to use the images of their precious dogs. Thank you Leah Rosenfeld (Chocolate Lab Rosie), Erin Strout (Mixed Breed Saylor), Eileen Anderson (Mixed Breed Lewis), Stella Riskin (Mixed Breed Bodie), Heather Brown (Bichon Frise Brian), and Travis Agnew (English Mastiff Baldur). I am grateful to graphic designer Heather Brown of Cultural Sponge for creating a cover that makes me so happy.

Thank you to the *Arizona Daily Sun* for letting me write the animal column since 2008. I appreciate so much the managing editors of the paper—Chris Etling and the late Randy Wilson—for keeping the newspaper going strong when many towns our size have lost their local papers. I appreciate all the people who edited my column over the years, including Arlene Hittle, Svea Conrad, Larry Hendricks, and Sam McManis. Thanks also to Sheila Madrak, who was at the helm of the column before me and invited me to write a guest column for her one day when we were hanging out and talking about dogs at the local humane association where she worked.

Thank you to my editor, Eileen Anderson, whose huge improvements in my writing because of her knowledge of grammar, dogs, and publishing should be a cause for gratitude for anyone reading this book. Also, I so appreciate it that she laughs when I joke that I can't decide which name is better for her: 1) Eileen the Mighty Word Slayer or 2) Eileen the Mighty, Word Slayer. It's a joy and a riot of laughter to work together.

# About the Author

Karen B. London, PhD, is a Certified Professional Dog Trainer and a Certified Applied Animal Behaviorist. She received her BS in Biology from UCLA and her PhD in Zoology from the University of Wisconsin-Madison, where she studied the defensive behavior of neotropical social wasps, and a nesting association between two species of wasps.

She began working professionally with dogs in 1997, and has spent years working with clients in one-on-one consultations, teaching group classes, and giving seminars about dog training and canine ethology for trainers, veterinary and shelter staff, and the public.

She writes for Kinship.com, writes the animal column for the *Arizona Daily Sun,* and is an Adjunct Professor in the Department of Biological Sciences at Northern Arizona University, where she has enjoyed teaching tropical ecology and conservation field courses in Nicaragua and Costa Rica. She also taught a class for first year students about the importance of insects to society called "Sex, Bugs, and Rock 'n' Roll."

Karen and her husband live in Flagstaff, Arizona where they raised their two sons.

Instagram: @Karen.London.Dog.Behavior

Made in the USA
Monee, IL
08 November 2026

67971951R00083